MW01058932

Interdisciplinary Statistics

STATISTICAL ANALYSIS of GENE EXPRESSION MICROARRAY DATA

CHAPMAN & HALL/CRC

Interdisciplinary Statistics Series

Series editors: N. Keiding, B. Morgan, T. Speed, P. van der Heijden

AN INVARIANT APPROACH TO STATISTICAL ANALYSIS OF SHAPES	S. Lele and J. Richtsmeier
ASTROSTATISTICS	G. Babu and E. Feigelson
CLINICAL TRIALS IN ONCOLOGY SECOND EDITION	J. Crowley, S. Green, and J. Benedetti
DESIGN AND ANALYSIS OF QUALITY OF LIFE OF STUDIES IN CLINICAL TRIALS	Diane L. Fairclough
DYNAMICAL SEARCH	L. Pronzato, H. Wynn, and A. Zhigljavsky
GRAPHICAL ANALYSIS OF MULTI-RESPONSE DATA	K. Basford and J. Tukey
INTRODUCTION TO COMPUTATIONAL BIOLOGY: MAPS, SEQUENCES, AND GENOMES	M. Waterman
MARKOV CHAIN MONTE CARLO IN PRACTICE	W. Gilks, S. Richardson, and D. Spiegelhalter
STATISTICAL ANALYSIS OF GENE EXPRESSION MICROARRAY DATA	Terry Speed
STATISTICS FOR ENVIRONMENTAL BIOLOGY AND TOXICOLOGY	A. Bailer and W. Piegorsch

Interdisciplinary Statistics

STATISTICAL ANALYSIS of GENE EXPRESSION MICROARRAY DATA

Edited by

Terry Speed

A CRC Press Company

Boca Raton London New York Washington, D.C.

Library of Congress Cataloging-in-Publication Data

Statistical analysis of gene expression microarray data / edited by Terry Speed.
p. cm. -- (Interdisciplinary statistics)
Includes bibliographical references and index.
ISBN 1-58488-327-8 (alk. paper)
1. DNA microarrays--Statistical methods. 2. Gene expression--Statistical methods. I. Speed, T. P. II. Series.

QP624.5.D726 S73 2003
572.8′636--dc21 2002041298

This book contains information obtained from authentic and highly regarded sources. Reprinted material is quoted with permission, and sources are indicated. A wide variety of references are listed. Reasonable efforts have been made to publish reliable data and information, but the authors and the publisher cannot assume responsibility for the validity of all materials or for the consequences of their use.

Neither this book nor any part may be reproduced or transmitted in any form or by any means, electronic or mechanical, including photocopying, microfilming, and recording, or by any information storage or retrieval system, without prior permission in writing from the publisher.

All rights reserved. Authorization to photocopy items for internal or personal use, or the personal or internal use of specific clients, may be granted by CRC Press LLC, provided that $1.50 per page photocopied is paid directly to Copyright Clearance Center, 222 Rosewood Drive, Danvers, MA 01923 USA. The fee code for users of the Transactional Reporting Service is ISBN 1-58488-327-8/03/$0.00+$1.50. The fee is subject to change without notice. For organizations that have been granted a photocopy license by the CCC, a separate system of payment has been arranged.

The consent of CRC Press LLC does not extend to copying for general distribution, for promotion, for creating new works, or for resale. Specific permission must be obtained in writing from CRC Press LLC for such copying.

Direct all inquiries to CRC Press LLC, 2000 N.W. Corporate Blvd., Boca Raton, Florida 33431.

Trademark Notice: Product or corporate names may be trademarks or registered trademarks, and are used only for identification and explanation, without intent to infringe.

Visit the CRC Press Web site at www.crcpress.com

© 2003 by Chapman & Hall/CRC CRC Press LLC

No claim to original U.S. Government works
International Standard Book Number 1-58488-327-8
Library of Congress Card Number 2002041298
Printed in the United States of America 1 2 3 4 5 6 7 8 9 0
Printed on acid-free paper

Dedication

We dedicate this book to:

My mother Zhaohua (Cheng Li)

My asleep father Ming-Fa (George C. Tseng)

My wife Lisa (Wing Hung Wong)

Shaw (Yee Hwa Yang)

Sally (Terry Speed)

My husband Jeff (Sandrine Dudoit)

My parents Mikhail and Nina (Jane Fridlyand)

My wife Tania (Hugh Chipman)

Contents

Contributors

Hugh Chipman
University of Waterloo
Department of Statistics and Actuarial Science
Waterloo, Ontario, Canada

Sandrine Dudoit
University of California
Division of Biostatistics, School of Public Health
Berkeley, California

Jane Fridlyand
University of California
Comprehensive Cancer Center
San Francisco, California

Trevor J. Hastie
Stanford University
Department of Statistics
Stanford, California

Cheng Li
Harvard School of Public Health
Department of Biostatistics
Boston, Massachusetts

Terry Speed
University of California
Department of Statistics
Berkeley, California
and
Walter and Eliza Hall Institute of Medical Research
Division of Genetics and Bioinformatics
Melbourne, Australia

Robert Tibshirani
Stanford University
Department of Health Research and Policy
Department of Statistics
Stanford, California

George C. Tseng
Harvard School of Public Health
Department of Biostatistics
Boston, Massachusetts

Wing Hung Wong
Harvard University
Department of Statistics
Department of Biostatistics
Cambridge, Massachusetts

Yee Hwa Yang
University of California
Division of Biostatistics
San Francisco, California

Preface

The field of microarray data analysis is less than a decade old, but it is already occupying the time and energies of a large and growing number of statisticians and others. It appears clear to us that large-scale gene expression studies are not a passing fashion, but are instead one aspect of a new mode of biological experimentation, one involving large-scale, high throughput assays. Themes here include parallel approaches to the collection of very large amounts of data (by biological standards), quite sophisticated instrumentation that needs understanding by statisticians, data where the systematic features are at least as important as the random ones, and a general sense that we are dealing more with industrial scale than the traditional single-investigator lab research, with data compiled in a notebook. Furthermore, this kind of research often involves many different kinds of data, including clinical, genetic, and molecular, as well as the basic assay data, and so topics of data integration and the use of databases readily arise.

Although the details of the technologies will undoubtedly change over time, the opportunities for serious statistical engagement will remain. We hope our readers will find this field as interesting as we do, and join us. More than enough datasets and problems are available to go around.

Acknowledgments

We thank the following people for collaborations, discussions, data, or help in reading drafts: Eric Schadt, Sven de Vos, Dan Tang, Nik Brown, Stanley Nelson, Jae K. Lee, Yaron Hakak, John Walker, Arindam Bhattacharjee, Katrin Wuennenburg-Stapleton, Matt Callow, David Kimelman, David Raible, Dave Lin, Jonathan Scolnick, Elva Diaz, John Ngai and his lab members, Matt Ritchie, Natalie Thorne, Gordon Smyth, David Bowtell, Patty Solomon, Gary Glonek, Julia Brettschneider, Yongchao Ge, Yun Zhao, Ming-Ying Leung, Leo Breiman, Joseph Costello, Marcel Dettling, Robert Gentleman, Ajay Jain, Mark van der Laan, Therese Sørlie, and Gavin Sherlock.

We also gratefully acknowledge the following grant support:

NIH 1R01HG02341-01, NSF DMS-0090166, and NSF DBI-9904701 (Cheng Li, George C. Tseng, and Wing Hung Wong)

NIH RO1 GM59506 (T.P.S.), NIH R01 MH61665 (Yee Hwa Yang)

CA89520 and CA64602 (Jane Fridlyand)

NIH R01-LM07609-01 (Sandrine Dudoit)

DMS-9803645 and NIH ROI-CA-72028-01 (Trevor J. Hastie), NSERC Canada (Hugh Chipman)

CHAPTER 1

Model-based analysis of oligonucleotide arrays and issues in cDNA microarray analysis

Cheng Li, George C. Tseng, and Wing Hung Wong

Abstract. This chapter describes the model-based analysis of oligonucleotide arrays, including expression index computation, outlier detection, and standard error applications, as well as issues in the analysis of cDNA array data such as normalization, handling of replicate arrays and spots, and hierarchical modeling of the data in detecting differentially expressed genes. Software implementing these analysis methods can be found at http://biosun1.harvard.edu/complab/.

1.1 Model-based analysis of oligonucleotide arrays

1.1.1 Background

Oligonucleotide expression array technology (Lockhart et al., 1996) has recently been adopted in many areas of biomedical research to measure the abundance of messenger ribonucleic acid (mRNA) transcripts for many genes simultaneously. As reviewed in (Lipshutz et al., 1999), 11 to 20 perfect match (PM) and mismatch (MM) probe pairs are used to interrogate each gene (Figure 1.1), and the simple or robust average of the PM–MM differences for all probe pairs in a probe set (called "Average Difference" or "Signal" (Affymetrix Inc., 2001a,b)) is used as the expression index for the target gene (We use the term "probe" to refer to the deoxyribonucleic acid (DNA) sequence immobilized on the solid substrate/array, and the term "target" for the DNA or RNA sequence from the sample being interrogated). Researchers rely on the expression indexes as the starting point for "high-level analysis" such as self-organizing maps (SOM) (Tamayo et al., 1999) or two-way clustering (Alon et al., 1999). Besides the original publications by Affymetrix scientists (Lockhart et al., 1996; Wodicka et al., 1997), researchers are also exploring alternative analysis approaches on "low-level" issues such as feature extraction, normalization, and computation of expression indexes (Irizarry et al., 2003; Schadt et al., 2001a,b).

1-58488-327-8/03/$0.00+$1.50
© 2003 by CRC Press LLC

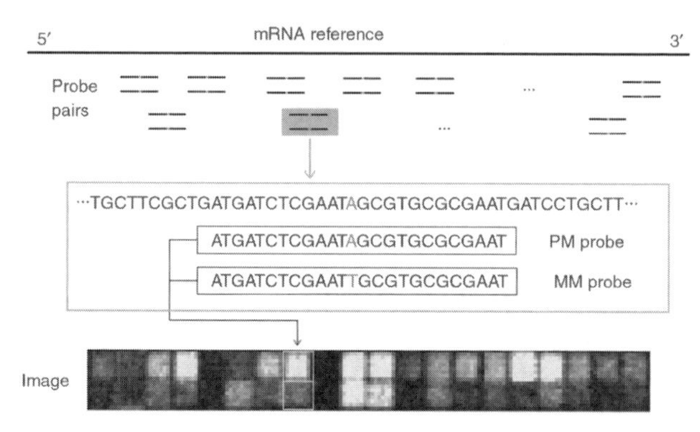

Figure 1.1 *Short 25-basepair oligonucleotides are selected from the gene of interest and synthesized directly onto the array. The intensities of perfect match (PM) and mismatch (MM) probes are used to infer the mRNA abundance of the target gene in a sample.*

The analysis of such experiments is nontrivial because of large data size and many levels of variation introduced at different stages of the experiments. The analysis is further complicated by the large differences that may exist among different probes used to interrogate the same gene. We have found that, even after making use of the control information provided by the MM probes, the information on expression level provided by the different probes for the same gene is still highly variable. We use a set of 21 Hu6800 arrays (Hofmann et al., 2002) to illustrate our discussion. This data set is typical, in terms of quality and sample size, of a data set from a single-laboratory experiment. We have applied the methodology to many sets of arrays from different laboratories and obtained similar results. Each of these 21 arrays contains more than 250,000 probe features and 7129 probe sets.

Figure 1.2 and Figure 1.3 show data for one probe set in the first 6 arrays (probe-level data is first normalized by the Invariant Set Normalization method (Li and Wong, 2001a)). This probe set (No. 6457) will be called probe set A hereafter. Considerable differences exist in the expression levels of this gene in the samples being interrogated because the between-array variation in PM–MM differences is substantial. More noteworthy is the dramatic variation among the PM–MM differences of the 20 probes that interrogate the transcript level. Analysis of variance of the PM–MM differences of this probe set in these 21 arrays shows that the variation due to probe effects is larger than the variation due to arrays. Specifically, mean square due to probes and arrays are 38,751,018 and 17,347,098 respectively. This is a general phenomenon: for the majority of the 7129 probe sets, the root mean square due to probes is five times or more of that due to arrays. Thus, it is clear that proper treatment of probe effects is an essential component of any approach to the analysis of such expression array data.

However, an attractive feature of high-density oligonucleotide arrays such as those produced by photolithography and ink-jet technology is the standardization of chip manufacturing and hybridization process. As a result, probe-specific biases, although significant, are highly reproducible and predictable, and their adverse effect can be

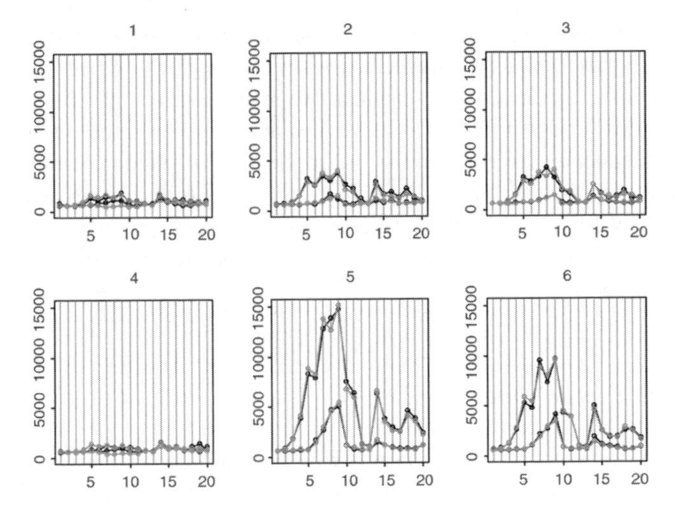

Figure 1.2 *Black curves are the PM and MM data of gene A in the first six arrays. Light curves are the fitted values of model (1.1). Probe pairs are labeled 1 to 20 on the x-axis.*

reduced by proper modeling and analysis methods. Here, we propose a statistical model for the probe-level data in order to account for probe-specific effects and develop model-based gene expression indexes.

In addition, human inspection and manual masking of image artifacts is currently very time-consuming and represents a limiting factor in large-scale expression profiling projects. We show that the goodness of fit to our model can be used to construct diagnostics for cross-hybridizing probes, contaminated array regions, and other image

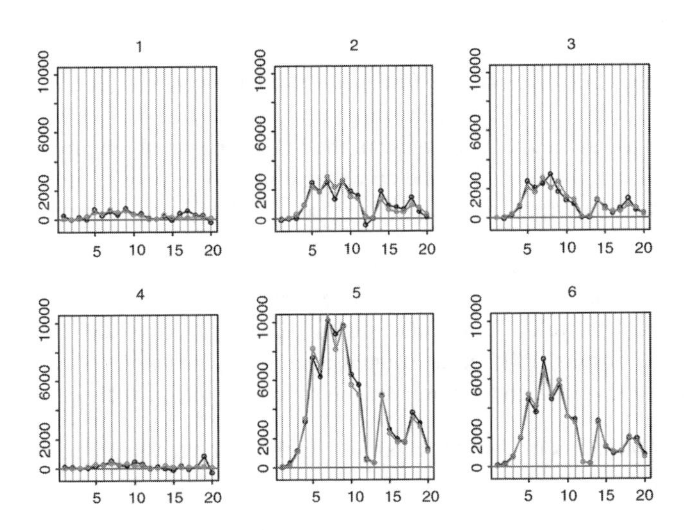

Figure 1.3 *Black curves are the PM–MM difference data of gene A in the first six arrays. Light curves are the fitted values of model (1.3).*

artifacts. We use the diagnostics to develop automated procedures for detecting and handling of all these artifacts. This method makes it possible to process and analyze a large number of arrays in a speedy manner. We investigate the stability of the probe-sensitivity index across different tissues types and the reproducibility of results in replicate experiments. We also discuss the application of standard errors (SE) in the downstream analysis, such as the confidence intervals of fold changes and assessment of the impact of SE on clustering results. A software package known as DNA-Chip Analyzer (dChip) is freely available at www.dchip.org for researchers to use these analysis methods.

1.1.2 Statistical model for a probe set

Suppose that a number ($I > 1$) of samples have been profiled in an experiment. Then, for any given gene, our task is to estimate the abundance level of its transcript in each of the samples. The expression-level estimates are constructed from the $2 \times I \times 20$ (assuming a probe set has 20 probe pairs) intensity values for the PM and MM probes corresponding to this gene. The estimation procedure is based on a model of how the probe intensity values respond to the changes of the gene expression level. Let us denote the expression index for the gene in sample i by θ_i (model-based expression index, MBEI). We assume that the intensity value of a probe will increase linearly as θ_i increases, but that the rate of increase will be different for different probes. We also assume that within the same probe pair, the PM intensity will increase at a higher rate than the MM intensity. We then have the following simple model:

$$\begin{aligned} MM_{ij} &= \upsilon_j + \theta_i \alpha_j + \varepsilon \\ PM_{ij} &= \upsilon_j + \theta_i \alpha_j + \theta_i \phi_j + \varepsilon. \end{aligned} \tag{1.1}$$

Here, PM_{ij} and MM_{ij} denote the PM and MM intensity values for array i and probe pair j for this gene, υ_j is the baseline response of the probe pair j due to nonspecific hybridization, α_j is the rate of increase of the MM response of the probe pair j, ϕ_j is the additional rate of increase in the corresponding PM response, and ε is a generic symbol for random error. The rates of increase are assumed to be nonnegative.

We fit Equation (1.1) to the $2 \times I \times 20$ data matrix for probe set A and Figure 1.2 shows the observed and fitted PM and MM intensities for the first six arrays. The model fits the data well. The residual sum of squares is only 1.03% of the sum of squares of the original PM and MM intensities. Thus, this model is able to capture the main relations between the observed intensities for different arrays and probes.

The model for individual probe responses implies an even simpler model for the PM–MM differences:

$$y_{ij} = PM_{ij} - MM_{ij} = \theta_i \phi_j + \varepsilon_{ij}. \tag{1.2}$$

In the rest of this section, our discussion will be focused on this PM–MM difference model. Feedback from collaborating biologists have indicated that currently there is a strong preference to base all computation directly on the differences between the PM and MM responses in a probe pair. Early experiments using a murine array with a large number of probes (more than 1000) per gene had shown that the average difference is linear to the true expression level in a certain dynamic range (Lockhart et al., 1996). There is also a computational advantage in reducing to differences first because the fitting of the full data is a more difficult numerical task. Thus, in this first attempt to implement model-based statistical inferences, we focus mainly on the analysis of PM–MM differences directly. It should be noted that the MM responses do contain information on the gene expression levels, and that this information can be better recovered by analyzing the PM and MM responses separately.

Equation (1.2), the model for the PM–MM differences, is identifiable only if we constrain it in some way. Here, we simply make the sum squares of ϕ's to be J (the number of probe pairs):

$$y_{ij} = PM_{ij} - MM_{ij} = \theta_i \phi_j + \varepsilon_{ij}, \sum \phi_j^2 = J, \ \varepsilon_{ij} \sim N\left(0, \sigma^2\right). \qquad (1.3)$$

Least square estimates for the parameters are carried out by iteratively fitting the set of θ's and ϕ's, regarding the other set as known. For comparison, we also perform least square fitting using the more standard additive model:

$$y_{ij} = \mu + \theta_i + \phi_j + \varepsilon_{ij}. \qquad (1.4)$$

Figure 1.4 presents the plots of residuals versus fitted values for these two models. The residuals of the additive model show a systematic pattern indicating lack of fit. The magnitude of the σ estimate of the multiplicative model (Equation 1.3) is much smaller than that of the additive model (Equation 1.4) (1075 vs. 2705). The explained energy (R^2, the ratio of sum of squares of predicted values and sum of squares of original data) is 98.08% and 87.85%, respectively, for the two models. The multiplicative model (Equation 1.3), with 40 parameters, is able to capture the relations among 420 data points (Figure 1.3).

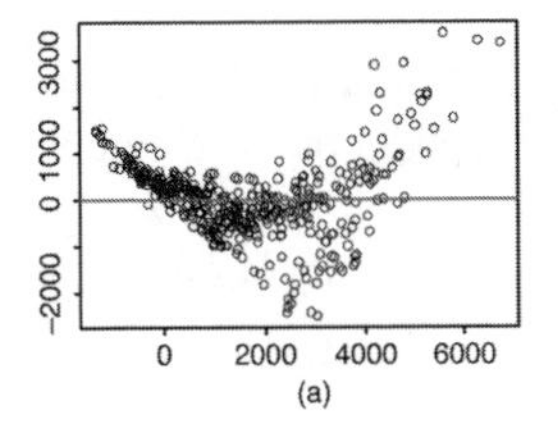

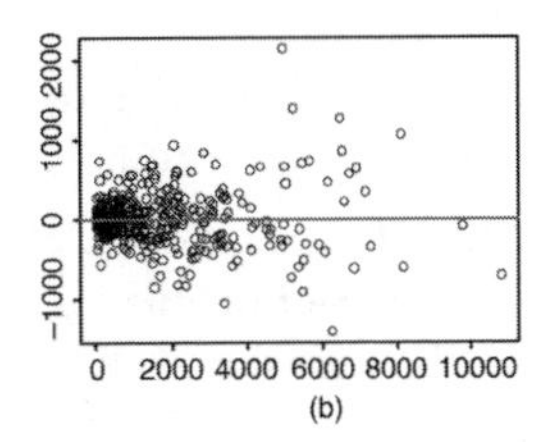

Figure 1.4 *Plot of residuals (y-axis) vs. fitted value (x-axis) for (a) additive model (Equation 1.4) and (b) multiplicative model (Equation 1.3).*

1.1.3 Conditional mean and standard error

Suppose for gene A, the ϕ's have been learned from a large number of arrays. We can then treat them as known constants and analyze the mean and variance of the expression index estimate. For a single array, Equation (1.3) becomes:

$$y_j = PM_{ij} - MM_{ij} = \theta\phi_j + \varepsilon_j. \tag{1.5}$$

Given the ϕ's, the linear least squares estimate for θ is

$$\widetilde{\theta} = \frac{\sum_j y_j \phi_j}{\sum_j \phi_j^2} = \frac{\sum_j y_j \phi_j}{J}, \; with \; E(\widetilde{\theta}) = \theta \; and \; Var(\widetilde{\theta}) = \sigma^2/J.$$

Hence, an approximate standard error for the least squares estimate can be computed:

$$Std\ Error(\widetilde{\theta}) = \sqrt{\widehat{\sigma}^2/J}, \; with \; \widehat{\sigma}^2 = \Big(\sum_j \big((fitted - observed)^2\big)/(J-1).$$

Similarly, when we regard the estimated θ's as fixed, we can calculate standard errors of ϕ's. These standard errors will play an important role in the outlier detection and probe selection discussed later. We note that the previous calculation is conditional in the sense that the ϕ's are regarded as known constants. This is valid if we have a large number of arrays to estimate them accurately, otherwise, the uncertainty in the estimation of these probe-specific parameters must be taken into account in the standard error computation.

1.1.4 Probe selection and automatic outlier and artifact detection

Conceptually, we can extend Equation (1.3) to model the response of a probe set to all genes in the sample:

$$y_{ij} = \theta_i^{(1)}\phi_j^{(1)} + \theta_i^{(2)}\phi_j^{(2)} + \theta_i^{(3)}\phi_j^{(3)} + \cdots + \theta_i^{(n)}\phi_j^{(n)} + \varepsilon_{ij}, \tag{1.6}$$

where $\theta_i^{(k)}$ is the expression level of gene k in array i, $\phi_j^{(k)}$ is the sensitivity of probe j to gene k, and n is the total number of different human genes (we do not consider complications such as alternative splicing here). Ideally, we want a probe set to be specific: If a probe set is intended to interrogate gene k, then only the $\phi_j^{(k)}$'s should be nonzero (thus sensitive) and all other $\phi_j^{(k)}$'s should be 0 (thus specific). In this case, the observed y_{ij} are specific signals coming from the target gene and Equation (1.6) is reduced to Equation (1.3), and the expression indexes $\theta_i^{(k)}$'s can be correctly estimated. We note that Equation (1.6) is formally a special case of the factor analysis model that is widely used in social sciences (Press, 1972).

Although Affymetrix has developed prediction rules to guide the selection of probe sequences with high specificity and sensitivity (Lockhart et al., 1996; Affymetrix Inc., 2002), inevitably there remain some probes hybridizing to one or more nontarget genes. We expect most cross-hybridizing genes to have expression patterns (in a large set of samples) different from that of the target gene, and different probes in a probe set to cross-hybridize to different nontarget genes. For a nontarget interfering gene k, the sensitivity indexes $\phi_j^{(k)}$'s are expected to be small except for one or two probes in the probe set. The mixed response of a probe set to target and nontarget genes suggests that probe selection (in the analysis stage) may enhance the specificity in estimating the expression levels of the target gene.

In the standard analysis (Wodicka et al., 1997), the mean and standard deviation of the PM–MM differences of a probe set in one array are computed after excluding the maximum and the minimum. If the difference of a probe pair is more than 3 standard deviations from the mean, the probe pair is marked as an outlier in this array and discarded in calculating average differences of both the baseline and the experiment array. One drawback to this approach is that a probe with a large response might well be the most informative, but it may be consistently discarded. Furthermore, if we want to compare many arrays at the same time, this method tends to exclude too many probes.

We exploit our model to detect and handle cross-hybridizing probes, image contamination and outliers from other causes. For a particular probe set, its 20 ϕ values constitute its "probe response pattern," and the model hypothesizes that the 20 differences in an array should follow this pattern and are scaled by the target gene's expression index (θ) in this array. The (conditional) standard error attached to a fitted θ is a good measure of how the 20 differences in the corresponding array conform to the probe response pattern. For example, in Figure 1.5b, array 4 is identified as an "array-outlier" because the estimated θ_4 has large standard error. Close examination of Figure 1.5a reveals that the probe responses in array 4 deviates from the consistent patterns shown in the other arrays. This could be due to various reasons including image artifacts (Figure 1.6). Because the probe responses in this array may affect the fitting of the probe response pattern, we exclude outlier arrays (identified by large standard errors) and use the remaining arrays to estimate the probe response pattern.

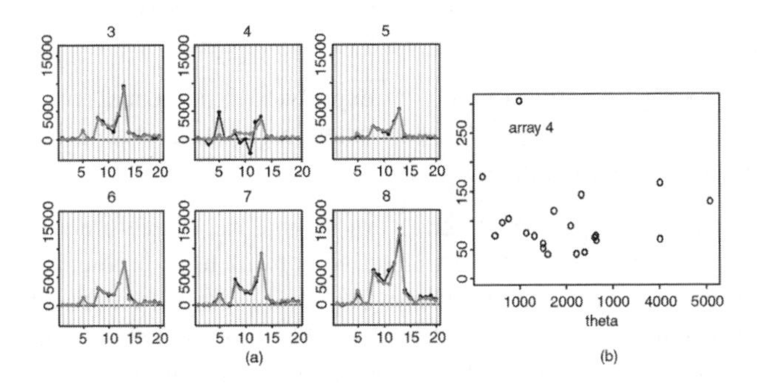

Figure 1.5 *(a) 6 arrays of probe set 1248. (b) Plot of standard error (SE, y-axis) vs. θ. The probe pattern (black curve) of array 4 is inconsistent with other arrays, leading to unsatisfactory fitted curve (light) and large standard errors of θ_4.*

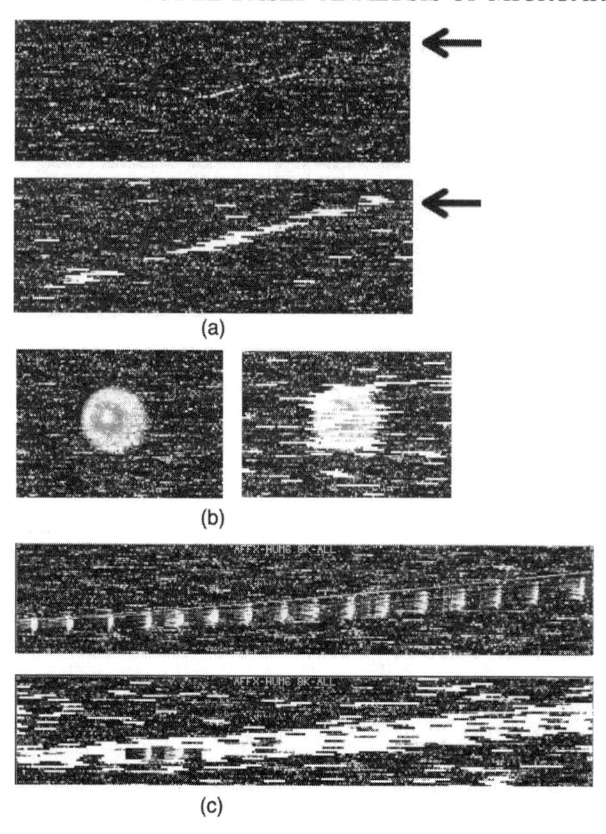

Figure 1.6 *(See color insert following page 114.) A long scratch contamination (indicated by arrow) is alleviated by automatic outlier exclusion along this scratch. (b,c) Regional clustering of array-outliers (white bars) indicates contaminated regions in the original images. These outliers are automatically detected and accommodated in the analysis. Note that some probe sets in the contaminated region are not marked as array-outlier, because the contamination contributes additively to PM and MM in a similar magnitude and thus cancels in the PM–MM differences, preserving the correct signals and probe response patterns.*

For an outlier array, we still compute its expression index conditional on the estimated probe response pattern, with the attached large standard error indicating poor reliability of this expression index.

In Equation (1.3), the roles of θ and ϕ are symmetric. Therefore we can use the conditional standard errors of the estimated ϕ_j's to identify problematic probes. In Figure 1.7a, probe 17 (indicated by arrow in several arrays) has peculiar behavior that is inconsistent with the rise and fall of other probes. This is probably due to the cross-hybridization of this probe to nontarget genes. Figure 1.7b shows that this nonspecific probe can be identified by the large standard error of ϕ_{17}. Finally, we must also consider a "single-outlier" which might be an image spike in one array, affecting just one PM–MM difference. Such a single-outlier (for example, y_{ij} in the data matrix)

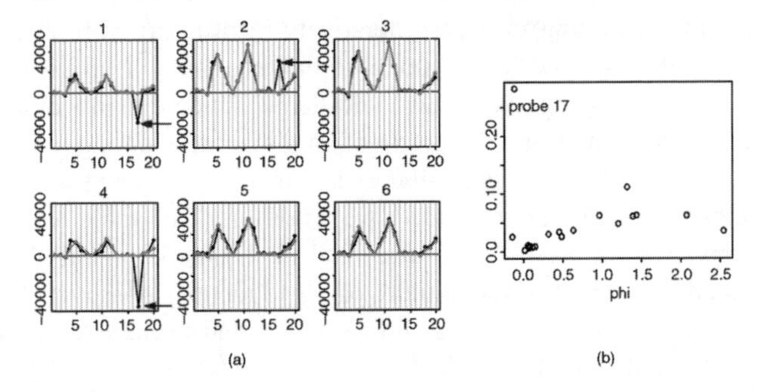

Figure 1.7 *(a) The probe 17 of probe set 1222 is not concordant with other probes (black arrows); (b) probe 17 is numerically identified by the large standard error of* ϕ_{17}.

may affect estimates of both θ_i and ϕ_j, and we can identify it by the large residual for this data point. Once identified, single-outliers are regarded as "missing data" in the model fitting; their values are imputed, so they do not affected the estimation of $\theta' s$ and $\phi' s$.

Besides array, probe, and single-outliers, several other undesirable artifacts occur in the data that we wish to handle. Figure 1.8a illustrates a responsive probe 12 amidst other nonresponsive probes. Generally, if the target gene exists in the samples, we expect more than one of the 20 probe pairs to respond at various sensitivities. In this case, it is most likely that the target gene is not present in any samples, and that the large response by probe 12 is due to cross-hybridization to nontarget genes. Although the model fits well in this case (99.83% variance explained), it is prudent to exclude this probe because of its unusually high leverage. Figure 1.8b shows that such a probe can be automatically identified through its large ϕ value. We classify a probe as "high-leverage" and exclude it during the fitting of the model, if it contributes more than 80% to the sum of squares of the ϕ_j's in the probe set. A similar procedure is used to identify high-leverage arrays.

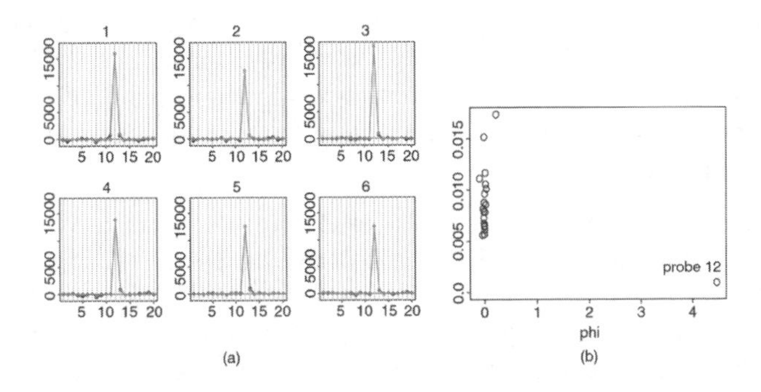

Figure 1.8 *(a) Probe set 3562 has a single high-leverage probe 12, and the fitted light curves almost coincide with the black data curve. (b)* ϕ_{12} *is large compared to other* ϕ*'s close-to-zero value. Note that Affymetrix's superscoring method works here by consistently excluding this probe across arrays.*

To implement the preceding ideas, we iteratively identify array, probe, and single-outliers. Specifically, we first fit the model to the data table (I arrays by J probes) of one probe set, identifying θ's (arrays) with large standard error (more than three times as large as the median standard error of all θ's) or dominating magnitude (θ^2 is more than 80% of sum of squares of all θ's), and mark these arrays as array-outliers. Next, with these array-outliers excluded, we work on a data table with fewer rows (discarding rows for the excluded arrays), and fit the model again. This time, we inspect the standard errors and magnitudes of ϕ's to identify possible probe-outliers. If a ϕ is negative, we also regard it as probe-outlier and exclude the corresponding probe. In effect, the data table shrinks in columns and we fit the model again to this new data table.

Note that, although some arrays and probes may not be used in fitting the model, we still can regress the data in one array (excluding probes not used in model fitting) against the estimated probe response pattern (ϕ's) to get estimate of expression levels (θ's) of the excluded arrays, and similar for the excluded probes. After probe-outliers are excluded, we evaluate all arrays for outliers again, and compare them to the set of array-outliers in the previous round to see if any change occurs. This procedure is repeated until the set of probe-outliers and array-outliers do not change any more. (In some cases, they may cycle between a small number of slightly different outlier sets.) Along the iteration, we will also identify some single data point outliers with large residuals and mark them as missing data when fitting the model. In general, five to ten iterations will lead to a converged set of outliers.

1.1.5 Model-fitting summary

We apply this model-based analysis to all the 7129 probe sets of the 21 Hu6800 arrays. As illustrated in Figure 1.6, image contamination can be handled automatically, by reasonably marking array and single-outliers and excluding them from model fitting. Such contamination would lead to incorrect expression and fold change calculation if left unattended in the data. As a quality control step in the experiment and analysis, arrays with a large number of array and single-outliers deserve further attention (for example, check if images or samples have been contaminated). The model automatically excludes these "outlier-arrays" from model fitting to avoid influence of these arrays on good ones, and attaches large standard errors to the expression indexes of contaminated probe sets.

Figure 1.9a demonstrates that, for 60.2% of the probe sets, we use more than half of the probes to fit the model. Figure 1.9b demonstrates that the explained energy is high (R^2 greater than 80%) for 63.3% of the probe sets. To investigate the reason for the low probe usage and poor R^2 for some probe sets, we examine the relation among probe usage, explained energy, and the presence percentage (percentage of arrays where a probe set is called "Present" by GeneChip, Figure 1.9c). Figure 1.10a shows that high presence percentage usually leads to high probe usage. Figure 1.10b demonstrates that when a gene is present in many arrays, the explained energy of the corresponding probe set tends to be high. Clearly, when a gene is absent in most

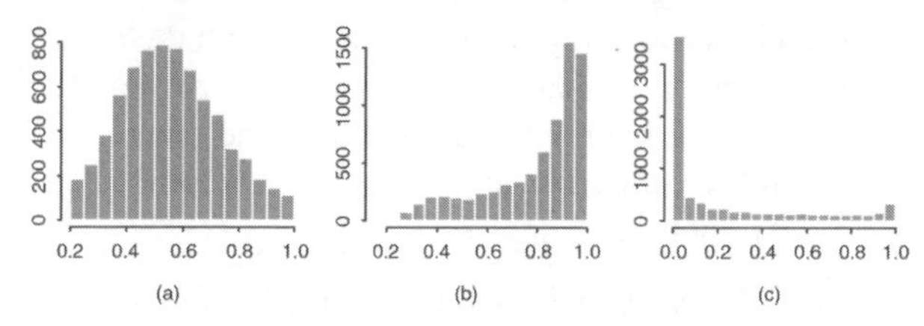

Figure 1.9 *(a) Histograms of percent of probes used; (b) explained energy; and (c) presence percentage for all 7129 probe sets. As observed in (c), most genes are only present in a few arrays.*

arrays, the variation in the observed data is mostly due to the noise term and one should not expect the model to explain a large fraction of this variation. In this case, the data does not include enough information to determine the ϕ's, but the θ's are still correctly estimated to be close to zero.

1.1.6 Probe-sensitivity indexes are stable across tissues types

In practice, a researcher hybridizes tissue or cell line samples corresponding to different treatments or conditions to a batch of arrays. Ideally, the probe-sensitivity index (ϕ) should be independent of the tissue type. This condition, however, may not hold for those probes that have cross-hybridization affinity to nontarget genes. Nevertheless, assuming that a nontarget gene cross-hybridizes only to a few probes of a probe set,

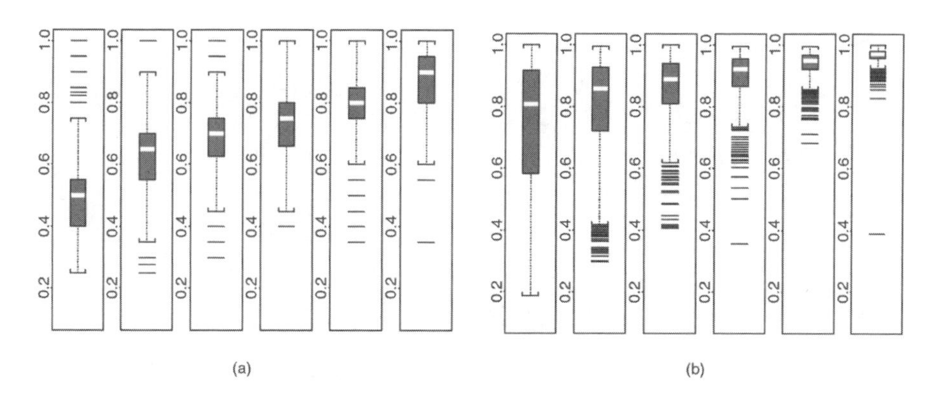

Figure 1.10 *Boxplots of probe usage (a) and explained energy (b) stratified by presence percentage (the number of presences of a gene in the 21 arrays and the subpopulation size for the 6 boxplots are: 0–3, 4365; 4–7, 817; 8–11, 567; 12–15, 520; 16–19, 518; and 20–21, 342). When presence percentage is high, the excluded probes tend to be cross-hybridizing probes; when presence percentage is low, PM–MM differences fluctuating around zero may result in many negative ϕ estimates and exclusion of the corresponding probes. As more arrays enter the database, we may reuse these probes if they respond positively to target expressions. The more arrays in which a target gene is present, the higher the explained energy.*

and its expression levels across arrays do not correlate with the target gene, the iterative probe-excluding procedure may be able to exclude cross-hybridizing probes, regardless of the tissue type hybridized. In addition, the relative probe sensitivity indexes of the good probes called by the model are likely to be similar across array sets hybridizing to different tissue samples.

Stability of probe-sensitivity index is studied using 226 HU6800 arrays. We apply Equation (1.3) independently to six sets of HU6800 arrays (21 leukemia, lymphoma, and mantle cell samples (Hofmann et al., 2002), 20 prostate cancer cell lines, 17 brain tumor samples, 55 cancer cell lines (Staunton et al., 2001), 58 brain samples (Hakak et al., 2001), and 55 lung tumor samples). Figure 1.11a shows the ϕ's fitted for probe set A in the six array sets. The ϕ patterns resemble each other greatly, showing the probe-sensitivity index is an inherent property of these non-cross-hybridizing probes and can be consistently identified from different sets of arrays. Figure 1.11b depicts

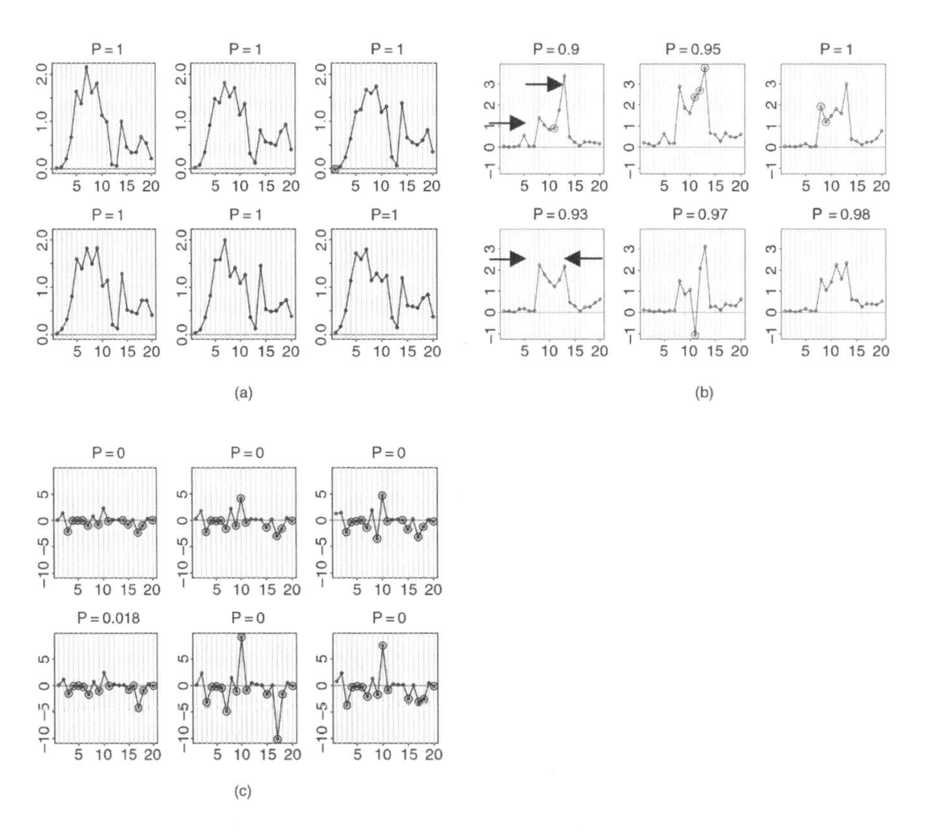

Figure 1.11 *ϕ's estimated for probe set 6457 (a), 1248 (b), and 6571 (c) in six array sets (shown in six panels); ϕ values (constrained to have sum square equal to number of non-outlier probes used in each array set) are on the y-axis, and probe pairs are labelled 1 to 20 on the x-axis. The title of each panel indicates the proportion of arrays called "Present" for the target gene in the array set. Large circles represent identified "probe-outliers" by negativity or large standard error of ϕ.*

the ϕ patterns for another probe set. Noteworthy is that the probe 11 in array set 5 is likely to be cross-hybridizing, making its relative strength (here, MM is consistently larger than PM and this leads to a negative ϕ) dissimilar to the probe 11 in other array sets. The model identifies this probe as a "probe-outlier" only for array set 5 and excludes it when calculating model-based expression indexes (MBEI) for array set 5.

In Figure 1.11a and b, the target gene is present in most samples of all array sets. For a probe set where the target gene is mostly absent throughout samples (Figure 1.11c), many probes are identified as "probe-outliers" because of their negative indexes. Here, we cannot obtain correct probe sensitivity indexes because of the absence of the target gene. Nevertheless, the PM–MM values for these probes are random fluctuations around zero, leading to correct expression index close to zero. If the target gene is expressed in the samples hybridizing to a future array set, the correct probe sensitivity indexes will be recovered and these probes will be used for expression calculation.

Occasionally, a responsive probe set may give rise to very different ϕ estimates in two array sets. In Figure 1.11b, probe 8 and 13 have different relative responses in array set 1 and 4 (indicated by arrows), leading to different probe response patterns. This might be due to the possibility that the probes in this probe set are differentially cross-hybridized in different array sets, or that the same probe in different batches of arrays may systematically behave differently. Identification and flagging such probe sets is desirable and essential if we want to compare arrays hybridized to different tissue samples.

Figure 1.12 illustrates the boxplots of average pairwise correlations of ϕ's between two array sets, stratified by average lower presence proportion in two array sets. In general, when a gene is present in many samples of two array sets, the ϕ patterns estimated from the two array sets are highly similar. This is because the target gene's presence in many arrays of an array set allows the probe-sensitivity indexes to be estimated accurately.

1.1.7 MBEI reduces variability for low expression estimates

The array set 5 has 29 pairs of replicate arrays (Hakak et al., 2001). Each pair consists of two arrays hybridizing to samples replicated at total mRNA level (the total mRNA sample is split and then amplified and labeled separately, and hybridized to two different arrays). The differences between the expression values of the two replicate arrays in a pair are due to the variation introduced in experimental steps after the split, the array manufacturing difference, and analytical methods such as normalization and expression calculation. This difference provides a lower bound of biological variation that can be detected between two independently amplified samples and serves as a good measure for comparing different analytical methods.

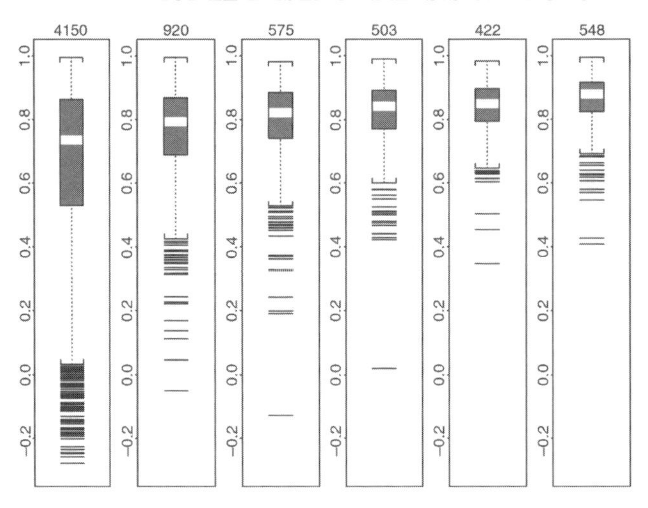

Figure 1.12 *Boxplots of average pairwise correlations of ϕ's between two array sets, stratified by average lower presence proportion in two array sets (the presence proportion of a probe set is the proportion of arrays in an array set where the target gene is called "Present" by GeneChip's algorithm). The average is taken over C(6, 2) = 15 pairwise comparisons of two array sets for each probe set, and the correlation is calculated using probes that are not identified as outlier in both array sets. The range of the average lower presence proportion for the six boxplots are: (0, 0.17), (0.17, 0.34), (0.34, 0.51), (0.51, 0.68), (0.68, 0.85), (0.85, 1). The title of each boxplot is the number of probe sets used in this boxplot. Eleven probe sets with too few non-outlier probes to calculate ϕ correlation for all the 15 comparisons are not included in the boxplots. The average lower presence proportion and average pairwise correlation for probe sets in Figure 1.11 are 1, 0.95 (a); 0.93, 0.94 (b); and 0, 0.86 (c).*

The agreement of MBEI between two replicate arrays is shown in Figure 1.13a. For comparison, we also use the method in (Wodicka et al., 1997) to calculate average differences (AD) for all probe sets and plot them in Figure 1.13b (AD is based on normalized probe values by the invariant set normalization method (Li and Wong, 2001b). Also, note that GeneChip software excludes probes whose PM–MM difference is outside 3 standard deviations of all probe differences in either of the two comparing arrays in the comparison; here, because we are comparing multiple arrays at the same time, when calculating ADs a probe is excluded if its difference is outlier in the previous sense in any of the arrays, until a minimum of five probes is reached where all five probes will be used. Both the MBEI and the AD method yield some expression values differing by more than a factor of two, especially for genes at low expression level. This might be explained by the relatively larger amplification variation for lowly expressed genes, given a constant success rate of amplifying a sequence by a certain fold.

Researchers often use a "log ratio" between expression values of a gene in two arrays as the criteria to identify differentially expressed genes. Between duplicate arrays, we expect these "log ratios" of expression values based on a good expression index

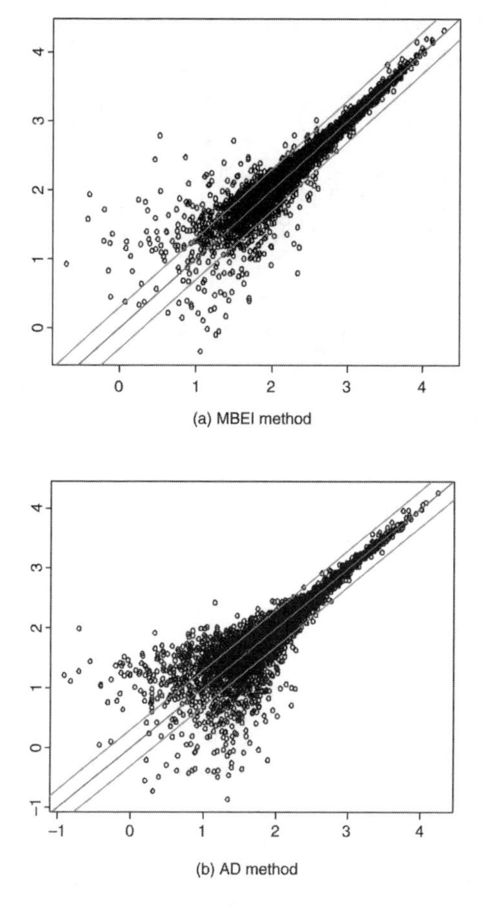

Figure 1.13 *Log (base 10) expression indexes of a pair of replicate arrays (array 1 and 2 of array set 5) for the MBEI method (a) and the AD method (b). Only 6695 (a) and 4696 (b) probe sets with positive values in both arrays are used. The center line is* y = x, *and the flanking lines indicate the difference of a factor of two.*

(AD or MBEI) to be close to zero. Thus, for every probe set, we calculate its average absolute log (base 10) ratio of 29 pairs of duplicates as a statistic to compare the variation between duplicates by their expression levels using the AD or the MBEI method. Figure 1.14 presents the result of the comparison. The average absolute log ratio distribution of the MBEI method is significantly lower than that of the AD method when expression level is low (thus, probe sets have low presence proportion across arrays). As the expression level becomes higher (when target gene of a probe set becomes present in more arrays), the AD method assumes a rapid performance improvement, approaching the level of the MBEI method. The result suggests that the MBEI method is able to extend the detection limit of reliable expressions to a lower level of mRNA concentration.

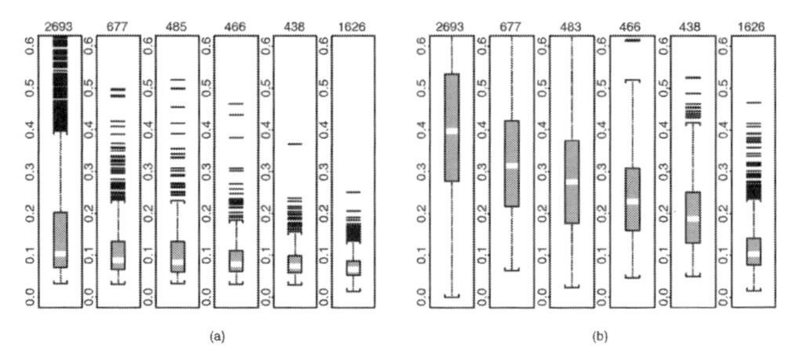

Figure 1.14 *Boxplots of average absolute log (base 10) ratios between replicate arrays stratified by presence proportion for (a) MBEI method, (b) AD method. The number of presence calls for a probe set in the 58 arrays for the 6 boxplots are: 0-9, 10-19, 20-29, 30-39, 40-49, 50-58. The title of each boxplot is the number of probe sets used for the boxplot. The average is taken over the 29 replicate pairs. Log ratios are not calculated for negative expression values or expression values identified as "array-outlier" by MBEI method in either array of a replicate pair, and are not used to calculate the average; 744 probe sets are not included because their average absolute log ratios cannot be calculated for all the 29 pairs using either method.*

1.1.8 Confidence interval for fold change

After obtaining expression indexes using AD or MBEI, fold changes can be calculated between two arrays for every gene and be used to identify differentially expressed genes. Usually, low or negative expressions are truncated to a small number before calculating fold changes, and GeneChip also cautions against using fold changes when baseline expression is absent.

The availability of standard errors for the model-based expression indexes allows us to obtain confidence intervals for fold changes. Suppose $\widehat{\theta}_1 \sim N(\theta_1, \delta_1^2)$, $\widehat{\theta}_2 \sim N(\theta_2, \delta_2^2)$, where θ_1 and θ_2 are the real expression levels in the two samples, and $\widehat{\theta}_1$ and $\widehat{\theta}_2$ are the model-based estimates of expression levels. We substitute the model-based standard errors for δ_1 and δ_2. Letting $r = \theta_1 / \theta_2$ be the real fold change, then the inference on r can be based on the quantity $Q = \frac{(\widehat{\theta}_1 - r\widehat{\theta}_2)^2}{\delta_1^2 + \delta_2^2 r^2}$.

It can be shown that Q has a χ^2 distribution with one degree of freedom irrespective of the values of θ_1 and θ_2 (Wallace, 1988). Thus, Q is a pivotal quantity involving r. We can use Q to construct fixed level tests and to invert them to obtain confidence intervals (CI) for fold changes (Cox and Hinkley, 1974).

Table 1.1 presents the estimated expression indexes (with standard errors) in two arrays and the 90% confidence intervals of the fold changes for 14 genes. Although all genes have similar estimated fold changes, the confidence intervals are very different. For example, gene 1 has fold change 2.47 and a tight confidence interval (2.07, 3.03). In contrast, gene 11 has a similar fold change of 2.49 but a much wider confidence interval (0.96, 18.18). Thus, the fold change around 2.5 for gene 11 is not as trustworthy as that for gene 1. Further examination reveals that this is due to the large standard errors relative to the expression indexes for gene 11. This agrees with the intuition

Table 1.1 *Using expression levels and associated standard errors to determine confidence intervals of fold changes.*

Gene	MBEI 1	SE 1	MBEI 2	SE 2	Fold change	Lower CB	Upper CB
1	860	42	348	36	2.47	2.07	3.03
2	406	31	164	44	2.47	1.67	4.49
3	284	29	115	18	2.48	1.84	3.48
4	46	64	19	85	2.48	0	∞
5	225	57	91	36	2.48	1.18	7.49
6	247	51	100	20	2.48	1.51	4.02
7	50	22	20	24	2.48	0.49	∞
8	276	19	111	36	2.48	1.59	5.35
9	436	33	175	21	2.49	1.99	3.19
10	76	18	30	18	2.49	1.07	86.17
11	81	25	32	17	2.49	0.96	18.18
12	182	42	73	28	2.49	1.25	7.12
13	1122	100	450	63	2.49	1.92	3.35
14	168	41	67	30	2.49	1.18	9.82

that when one or both expression levels are close to zero for one gene, the fold change cannot be estimated with much accuracy. In addition, when image contamination results in unreliable expression values with large standard errors, the fold changes calculated using these expression values are attached with wide CIs. In this manner, the measurement accuracy of expression values propagates to the estimation of fold changes. In practice, we find it useful to sort genes by the lower confidence bound ("Lower CB" in Table 1.1), which is a conservative estimate of the fold change.

1.1.9 Standard errors help to assess clustering results

Clustering analysis is a popular method to analyze the data of a series of microarrays (Eisen et al., 1998; Tamayo et al., 1999). If two genes are co-regulated at the transcription level, their expression values across samples are likely to be correlated. Clustering algorithms use these correlations (or the monotone transformation of correlations) to cluster co-regulated genes together; however, the correlation based on the estimated expression levels may be different than that based on the real but unobserved expression levels. Also the commonly used hierarchical clustering algorithm is an irreversible process: Once two genes or nodes are merged, they will stay together, even if later on there is good reason to adjust the previous clustering. Thus, the reliability of clusters must be assessed.

A global way of using standard error in hierarchical clustering is to resample or bootstrap (Efron and Tibshirani, 1993; Kerr and Churchill, 2001a) the whole "gene by sample" data matrix and redo the clustering, then investigate the overall properties emerging from this repertoire of clustering trees. In (Bittner et al., 2000), the data matrix coming from complementary deoxyribonucleic acid (cDNA) microarray experiments is resampled using the estimated variation derived from the median standard deviation of log-ratios for a gene across samples. Now we have standard errors for all data points, therefore, we can resample each expression value from a normal distribution with mean equal to the estimated expression value and standard deviation equal to the attached standard error.

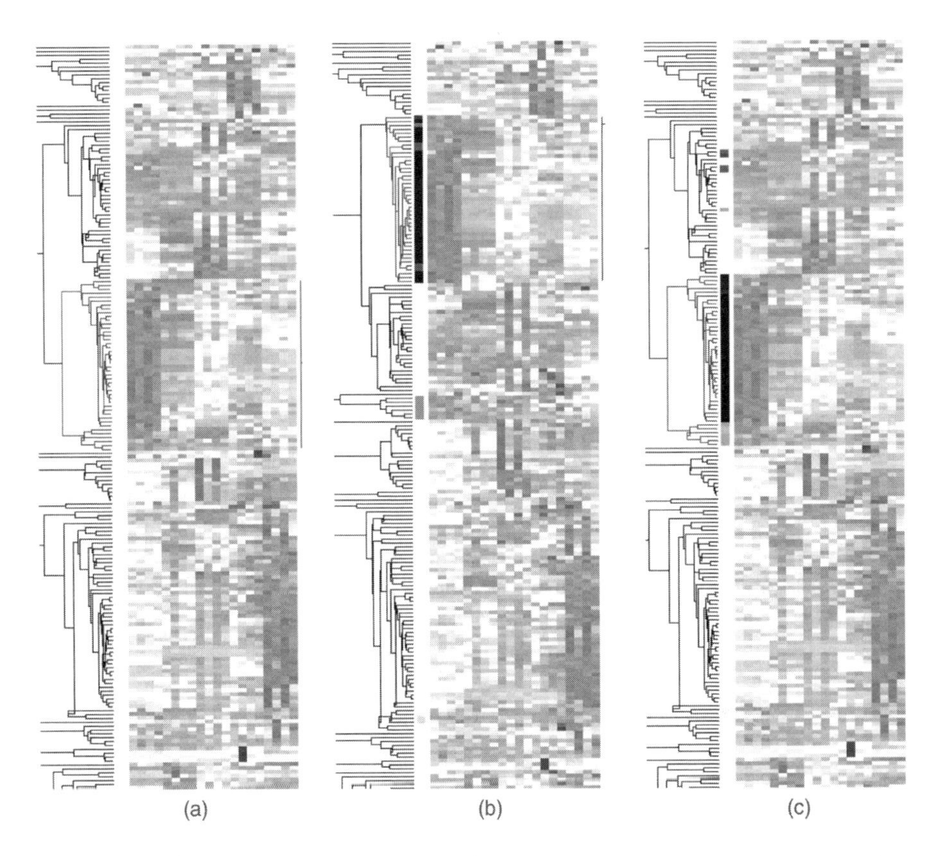

Figure 1.15 *(See color insert following page 114.) (a) 225 filtered genes are clustered based on their expression profiles across the 20 samples. Each gene's expression values are standardized to have mean zero and standard deviation one across 20 samples. Blue represents lower expression level and red higher expression level. We may be particularly interested in the gene branch colored in blue. (b) The clustering tree after a particular resampling. Although the original "blue" genes are scattered to various places, we can still determine where the original cluster is, using the criteria described in the text. (c) After resampling 30 times, the reliability of the genes belonging to the original cluster is indicated by the vertical gray-scale bar on the right of the clustering tree.*

Figure 1.15a depicts a hierarchical clustering tree of 225 selected genes with presence proportion >0.5 and coefficient of variation (standard deviation/mean) > 0.7 across the 20 samples in array set 2. In trying to interpret this tree, we may be interested in the gene cluster colored in blue and the reliability of the gene members belonging to this cluster. The whole data matrix is resampled, and the clustering is performed again (Figure 1.15b). We notice that some "blue" genes (genes in the original cluster are colored blue) are clustered with other non-blue genes, and some non-blue genes are mixed into the main body of blue genes. After each resampling, we identify a cluster that contains more than 80% of all the blue genes but as few non-blue genes as possible (measured in percentage of all genes in this new cluster). This cluster is considered to be the cluster that corresponds to the original one in Figure 1.15a. In Figure 1.15b, the root node of the "corresponding cluster" is marked with small horizontal line intersecting the vertical line representing the range of the cluster on the right of the clustering picture. Then, for each of all the 225 genes, if it belongs to this "corresponding cluster," we increase its "in-cluster" count by 1. After resampling 30 times, the in-cluster counts are indicated in gray-scale on the right side of the original clustering tree (Figure 1.15c), with black color represent 30 and white color representing 0. A high "in-cluster" count indicates a gene "remains" in the original cluster in most of the resampled clustering trees.

We can see from Figure 1.15c that most genes in the original cluster are reliable members, while a few genes at the bottom of the cluster are not (in fact they are merged into the original cluster in the last step). Interestingly, some genes in the upper part do not belong to the original cluster, but cluster with the "corresponding clusters" during resampling many times and have gray "in-cluster" marks. These genes may be related to the original cluster in some way. In summary, this method can help us to distinguish reliable and unreliable gene members of a gene cluster, as well as draw our attention to related genes originally clustered somewhere else because of the accidental nature of the hierarchical clustering.

1.1.10 Conclusion

We have proposed a statistical model at the probe level to compute expression indexes for oligonucleotide expression array data. Based on this model, we are able to address several important analysis issues that are difficult to handle using previous approaches, such as accounting for individual probe-specific effects, and automatic detection and handling of outliers and image artifacts. A software package DNA-Chip Analyzer (dChip) is available at www.dchip.org to perform normalization, calculation of MBEI, computation of confidence intervals of fold changes, and hierarchical clustering with resampling. Our experience is that more than ten arrays are appropriate for model fitting, outlier detection, and MBEI calculation.

Researchers are actively exploring and comparing the different expression index computation methods (Irizarry et al., 2003; Lemon et al., 2002; Holder et al., 2001; Naef et al., 2001; Zhou and Abagyan, 2002; Affymetrix Inc., 2001b; Zhang et al., 2002).

We believe as more validation data are available, these low-level analysis methods will no doubt be improved to best utilize the great potential of the oligonucleotide microarray.

1.2 Issues in cDNA microarray analysis

1.2.1 Background

Besides Affymetrix oligonucleotide expression array, cDNA microarray is another popular high-throughput technology for the global monitoring of gene expressions. As introduced in (Brown and Botstein, 1999), a library of thousands of distinct cDNA clones were first spotted on a microarray slide. Then, mRNA samples of two tissues were extracted, separately reverse transcribed into cDNA and labeled with differential colors (Cy3 and Cy5). The mixture of labeled cDNA were co-hybridized onto the microarray, competing to bind to their complementary cDNA. Finally, the slide is scanned at different wavelengths of a laser or by a charge-coupled device (CCD) camera to obtain numerical intensities of each dye.

Suppose the dye Cy3 is used for reference sample. Statistical analysis usually relies on the log-ratios of two dyes, $log(Cy5/Cy3)$ (or sometimes the ratio, $Cy5/Cy3$) to assess the expression levels. Some researchers use base 2 in the logarithm because the intensities usually range from 0 to 2^{16}; other researchers use base 10 for the ease of intuition. Using a different base, however, does not affect the conclusion of analysis. Since the first application of this technology in 1995 (Schena et al., 1995), many efforts have been made to address related statistical issues (Chen et al., 1997; Newton et al., 2000; Dudoit et al., 2002b; Kerr et al., 2001; Tseng et al., 2001). The huge data size and many variations introduced in different experimental stages complicate the analysis. Different statistical models and approaches usually have their advantages and drawbacks as well.

1.2.2 Issues in low level analysis

For convenience, we use the data set presented in Tseng et al. (2001) for illustration. In the first group of experiments, each slide had 125 *E. coli* genes multiply spotted (four spots per gene) on it while, in the second group, each slide had 4129 genes singly spotted. The first and second group of experiments will be called the 125-gene project and 4129-gene project, respectively, hereafter. In the 4129-gene project, two calibration and two comparative experiments are performed on the $E.\ coli$ genome to address several important issues in the data analysis and the study design of microarray experiments (Figure 1.16). In calibration experiments, the same sample is divided to label with two dyes while, in comparative experiments, two different samples are used for the two dyes. Different slides in the same experiment hybridize with the same pool of labeled cDNA onto different slides, and different experiments in the same project redo the whole experiment with the same pool of mRNA. We will use

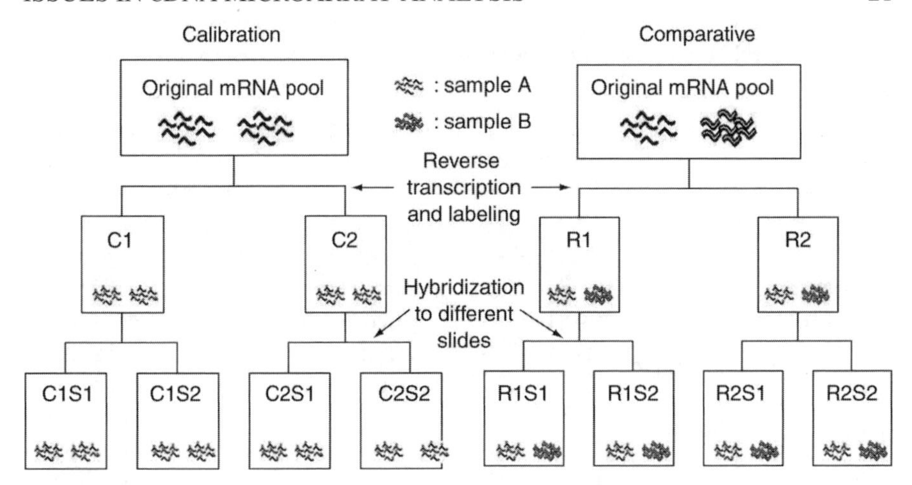

Figure 1.16 *(See color insert following page 114.) In calibration experiments, same samples are applied in the two dyes whereas, in comparative experiments, different samples are used with the two dyes. Replications are performed in experiment and slide stage. The 125-gene project has a similar design except for the quadruple spotting.*

C, R, and S to denote the calibration experiment, comparative (real) experiment, and slide, respectively, and the subsequent numbers to indicate sequence in both projects. For example, C2S2 indicates slide 2 in the second calibration experiment, and R1S2 means slide 2 in the first comparative experiment. The 125-gene project also has a similar experimental design.

In this data set, some calibration experiments used sample from *E. coli* grown in acetate, while the others are grown in glucose. The comparative experiments labeled mRNA from *E. coli* grown in acetate by Cy3 and mRNA from *E. coli* grown in glucose by Cy5. See Table 1.2 for details.

Several levels of replications are embedded in the design of the calibration experiments and the resulting data provide information on the relative importance of variations due to spots, labels, and slides. Based on this information, we formulate an approach to the analysis of comparative experiments. The main components are as follows:

1. Extract intensities from the scanned images of both dyes.
2. Detect and filter poor quality genes on a slide using measurement from multiple spots. This procedure is not applicable in singly spotted designs.
3. Perform slide-dependent nonlinear normalization of the log-ratios of the two channels.
4. Use hierarchical model-based analysis on normalized log-ratio scale, where assessment of the significance of gene effects are aided by statistical information obtained from calibration experiments, if they are available.

Table 1.2 *Experimental designs of 125-gene and 4129-gene project.*

		Slides in the experiment	Samples in Cy3	Samples in Cy5	Denhardt's solution
125 gene project	calibration	C1S1~C1S2	acetate	acetate	all slides
		C2S1~C2S4	glucose	glucose	none
		C3S1~C3S2	glucose	glucose	C3S1
		C4S1~C4S3	glucose	glucose	C4S1~C4S2
	comparative	R1S1~R1S2	acetate	glucose	all slides
		R2S1~R2S2	acetate	glucose	all slides
4129 gene project	calibration	C1S1~C1S2	acetate	acetate	all slides
		C2S1~C2S2	glucose	glucose	all slides
	comparative	R1S1~R1S2	acetate	glucose	all slides
		R2S1~R2S2	acetate	glucose	all slides

1.2.3 Image analysis

After hybridization and washing, slides are scanned by a laser or CCD scanner. The scanner then produces green Cy3 and red Cy5 16-bit TIFF image files. The intensity of each pixel in these images thus ranges from 0 to $2^{16}-1(=65,535)$. Image analysis in microarray experiments is a set of processes to extract meaningful intensities of each spot from the raw image for further analysis. The major components usually include:

1. **Locating the spots**. We need to first locate spot positions. Information like number of spots and prior rough positions are known from the arrayer (spotting machine), but an algorithm is needed to search for the exact location in the neighborhood. Usually some manual adjustments are needed.
2. **Segmentation**. This consists of deciding the shape of the spots and identifying foreground and background pixels. Some algorithms use only fixed diameters and round spot regions for each spot, some allow flexible diameters but use only round shapes, and some others allow both flexible diameters and irregular shapes. Background and foreground regions are then determined.
3. **Intensity extraction**. Local background intensities of each spot are then estimated and subtracted from the foreground intensities to account for cross-hybridization of nontarget genes and fluorescence emitted from other chemicals. Various statistics including mean intensities, median intensities, and standard deviation of the background and foreground of each dye are reported. Some of the statistics are used to provide intensity extraction, and others are used for quality control. The

spot summary information is very useful for the automation of quality filtering and further analysis.

Although the cDNA microarray experiment has been developed for several years, its image analysis is still an active area of research (Chen et al., 1997; Yang et al., 2002a; Buhler et al., 2002). It has some major difficulties. First of all, each cDNA clone usually contains several hundreds of pixels, and the locations and shapes of these spots may vary depending on the quality of the experiment. No fully automated algorithm can perfectly locate the spots and identify the regions on every slides, and most current software provides easy interface for manually adjusting spots that are wrongly identified by its algorithm. For some bad quality slides, these corrections may require tremendous labor. Second, a fast algorithm is necessary to deal with large datasets. Finally, many statistics are proposed to serve as quality indices. They are very useful in the case of misidentifying spot locations, local slide contamination, and poor spot quality. Some statistics are useful only to test some specific artifacts, however, and a good method to combine these statistics for correctly filtering all kinds of defect genes is not available yet.

Current available image analysis software include ArrayWorx, Dapple (Buhler, 2002), GenePix, ImaGene, ScanAlyze by Michael Eisen's lab (Eisen, 1999), Spot by Terry Speed's lab (Yang and Buckley, 2001), and UCSF Spot by Ajay Jain's lab (Jain et al., 2002). Some of them are free for academic use and others are commercial software.

1.2.4 Quality filtering

Quality filtering is usually implemented during the image analysis step using some quality indexes such as pixel-wise intensity variation or background intensities. Here, we discuss filtering using multiple spot replications.

Multiple spotting of target DNA on a slide provides a means to assess the quality of data for a gene on that slide (Lee et al., 2000). Suppose each gene is spotted p times on the slide ($p = 4$ in our 125-gene project). For each spot, a ratio of Cy3 and Cy5 intensity is first calculated as $m = Cy5/Cy3$. We denote by "CV" the coefficient of variation (i.e., standard deviation divided by mean) of the set of ratios $m_1, m_2, \ldots, m_p$ on the multiple spots. The quality of data on the expression level of each gene is inversely related to its CV. Figure 1.17 illustrates the CV versus mean intensity (average of Cy3 and Cy5 signals) on slides in the 125-gene project.

In Figure 1.17, we mark all genes having CV values larger than a threshold as poor quality data by a windowing procedure. For each gene, we construct a windowing subset by selecting 50 genes whose mean intensities are closest to this gene. If the CV of this gene is within the top 10% among genes in its windowing subset, then we regard the data on this gene as unreliable. The curves in Figure 1.17 depict the thresholds used to filter unreliable data. Data from both calibration and comparative experiments in the 125-gene project were filtered using this approach.

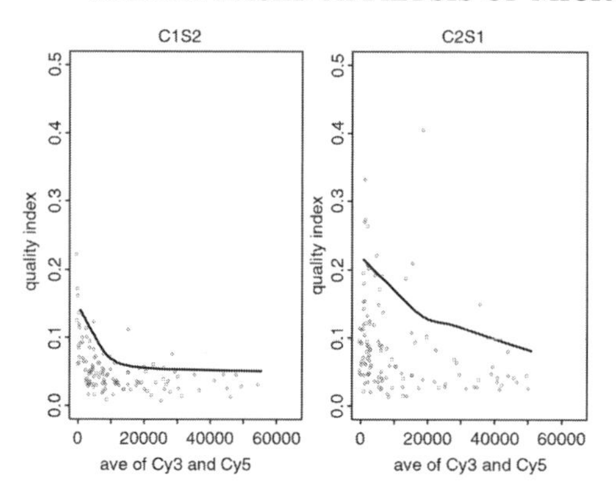

Figure 1.17 *Quality index (CV) vs. average intensity in 125-gene project. The curve indicates the 10th upper percentile in the moving windows containing 50 nearest genes. Genes with quality index (CV) larger than this curve will be filtered out. Only slides C1S2 and C2S1 are shown here. Genes with low CV have high agreement in duplicated spots, hence representing high experiment quality. Thus, slide C1S2 shows higher quality than slide C2S1.*

Following the convention of Dudoit, et al. (2002b), we draw the so-called MA–plot for initial investigation where $M = \log(Cy5/Cy3)$ representing log-ratio of two dyes and $A = (\log(Cy5) + \log(Cy3))/2$ the averaged logarithmic intensity. The plot is actually a 45-degree rotation and rescaling of log-intensity plot of Cy5 and Cy3. MA–plots of the remaining data of one calibration and one comparative slide after quality filtering are depicted in Figure 1.18.

Besides screening genes with unreliable data, the CV values can also be used to compare the quality of different slides and different experiments. For example, we found that C2S1∼4 have much poorer quality as compared with R1S1∼2, R2S1∼2 and

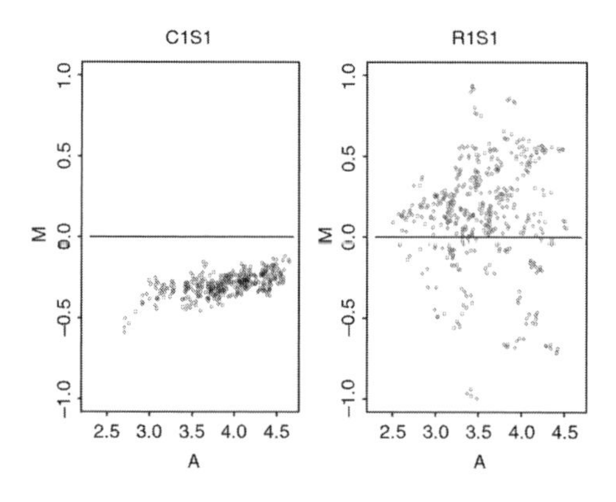

Figure 1.18 *Two MA–plots in the 125-gene project where M represents the log-ratio of two dyes and A represents the averaged logarithmic intensity. Only slides C1S1 and R1S1 are illustrated here.*

C1S1∼2 (Figure 1.17). Because Denhardt's solution was used in both comparative experiments and the first calibration experiment but not in the second calibration, we suspected this might be an explanation. To verify this, we performed the third and fourth calibration. It turned out that slides C3S1 and C4S1∼2 (with Denhardt's solution) have better quality compared with slides C3S2 and C4S3 (without Denhardt's solution) respectively in the same experiments (result not shown here). This confirms that using Denhardt's solution can greatly improve experiment quality. Thus, multiple spotting can provide useful information on data quality. It allows us to perform quality filtering (i.e., removing outlying spots, unreliable genes, or identifying problematic slides).

1.2.5 Normalization

Normalization in calibration

The most commonly used fluorescent dyes, Cy3 and Cy5, are relatively unstable. In addition, these dyes may differentially influence the incorporation efficiencies during labeling, have different quantum efficiencies, and are detected by the scanner with different efficiencies. The effect of these factors on intensity measurements is defined as the label effect, which is accounted for by the normalization curve in the following analysis.

To demonstrate the label effect, MA–plots in calibration experiments are illustrated in Figure 1.19. Because the two cDNA solutions were from the same pool of RNA in calibration, the scanner reading from the Cy3 channel should be identical to that from the Cy5 channel, if the label effects are negligible. In this ideal case, the MA–plot should scatter around the line $M = 0$. Figure 1.19 depicts the MA–plots after quality filtering in C1S1∼2 and C4S1∼2 in 125-gene project. It demonstrates that normalization is needed to account for the label effect. Another notable feature is that the normalization is slide-dependent. When the same batches of labeled cDNA were hybridized to a different slide, the MA–data showed a different correlation pattern (Figure 1.19, x vs. o). It suggests that no universal normalization curve exists. The following normalization procedures basically follow Dudoit et al. (2002b). First, we fit $\hat{M} = \hat{f}(A)$ on each slide in calibration experiment. The fitting can be done by the built-in "lowess" function in S-Plus (Venables and Ripley, 1998). Then, the normalized log-ratio is computed by $\tilde{M} = M - \hat{M}$.

Normalization in comparative experiments

In a comparative microarray experiment, two differentially expressed mRNA pools are separately labeled with Cy3 or Cy5 and co-hybridized to the same slide. As discussed previously, the label normalization function is nonlinear and slide-dependent. To perform label normalization in a comparative experiment, we have to identify on each slide a sufficient number of nondifferentially expressed genes and use them to construct a normalization curve.

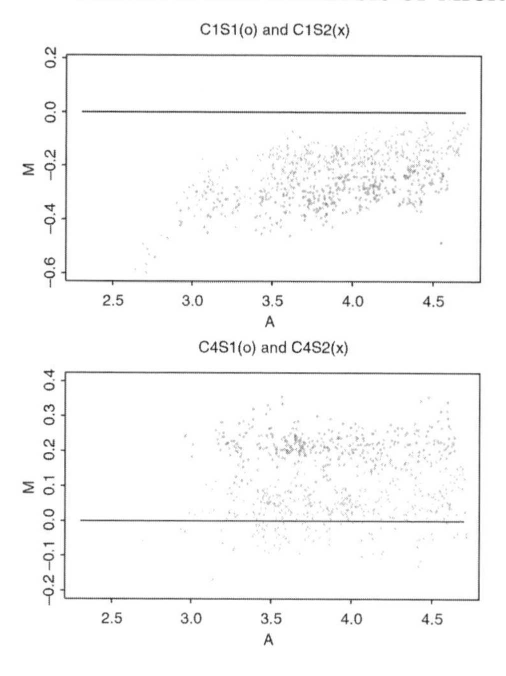

Figure 1.19 *(See color insert following page 114.) MA–plots of two slides in the same calibration experiment. The upper plot illustrates a different pattern of MA–plot on C1S1(o) and C1S2(x) in the 125-gene project. The lower MA–plot in calibration 4 also illustrates the same situation. Thus, the normalization curve is slide-dependent and should be estimated and applied within the same slide.*

One solution to this problem is to apply a set of predetermined "housekeeping" genes, which are biologically assumed to be nondifferentially expressed genes in the experiments. Note that, if the number of predetermined "housekeeping" genes is small or their intensities do not cover a range of different intensity levels, this approach may not provide a good fitting for nonlinear normalization curve. Also, the expression levels of "housekeeping" genes can exhibit natural variability.

Another approach is to use all the genes for the "lowess" curve fitting (Dudoit et al., 2002b). It requires the assumption that up- and down-regulated genes with similar average intensities (denoted A) roughly canceled out or otherwise most genes remain unchanged. This assumption is usually true in large genome studies.

Here, a rank-invariant selection scheme (Schadt et al., 2001b; Tseng et al., 2001) is introduced to improve the invariant gene selection. The ranks of Cy3 and Cy5 intensities of each gene on the slide are separately computed. For a given gene, if the ranks of Cy3 and Cy5 intensities differ by less than a threshold value d and the rank of the averaged intensity is not among the highest l ranks or lowest l ranks, this gene is classified as a nondifferentially expressed gene. In terms of a mathematical formula, we select the rank-invariant set

$$S = \{g : |Rank(Cy5_g) - Rank(Cy3_g)| < d \;\&\; l \\ l < Rank((Cy5_g + Cy3_g)/2) < G - l\} \tag{1.7}$$

A threshold value of 5 for both d and l was used in the 125-gene project. In the 4129-gene project, the larger number of genes allows us to use a more sophisticated iterative selection scheme (Tseng et al., 2001). In each iteration, the threshold for rank difference is determined by the number of selected genes (i.e., genes that have been selected in the last stage) multiplied by a predetermined percentage p. The threshold for rank averaged intensity l is only applied in the first iteration. The iteration stops when the remaining set of genes does not decrease after selection. We use $p = 0.02$ and $l = 25$ for the 4129-gene project. The result indicates that the iteration procedure helps to select a more conserved set of genes (result not shown here).

This method is based on the assumption that, if a gene is up-regulated, its intensity rank among one channel, for example, Cy5, should be significantly higher than the rank among the other and vice versa. This method may fail in some extreme cases where a majority of genes are up- (or down-) regulated to the same extent; however, if a large number of nondifferentially expressed genes exist, as in the case of most cDNA microarray experiment, this method will work well.

After selecting nondifferentially expressed genes and fitting the normalization curve as described previously, we extrapolate the normalization curve to normalize genes with extremely high or low intensities. The extrapolation is based on the 50 genes with highest and lowest averaged log-intensity rank in the selected set of nondifferentially expressed genes. Figure 1.20 depicts the extrapolated lowess curve in MA–plots in comparative experiments in 4129-gene project.

The within slide variation can also be large. Examples include areas of contamination, high background, or uneven cDNA hybridized on the slide surface. In experiments

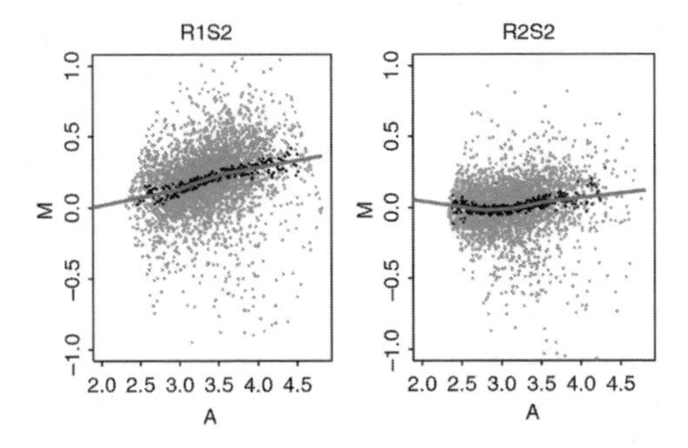

Figure 1.20 *(See color insert following page 114.) Normalization curves and MA–plots in comparative experiments in 4129-gene project. The darker points are genes of the rank-invariant set selected in an iterative manner (p = 0.02).*

using multiple pins to immobilize probe DNA, the pin-to-pin variation can be notable. We use a single pin in the 125-gene project and four pins in 4129-gene project. The pin-to-pin variation appears to be negligible in this dataset but was reported notable in Dudoit et al. (2002b). In that case, the normalization should be performed in a pin-to-pin manner.

1.2.6 *Some discussion about normalization*

We note that the preceding normalization procedure is nonlinear. The need for nonlinear normalization is also noted in Affymetrix oligonucleotide microarray analysis (Schadt et al., 2001). The normalized log-ratio $\tilde{M}$ can be expressed as $\log(K_A * Cy5/Cy3)$, which shows the multiplicative nature of the intensity-specific scaling factor. Many current softwares apply the ratio of mean or median logarithmic intensity of Cy3 to that of Cy5 as the constant scaling factor K. As we have observed in Figure 1.19, this is sometimes improper.

Another popular approach is to use the analysis of variance (ANOVA) model on logarithmic intensity scale that accounts for variations at each stage (dye, spot, slide, etc.) and their interaction terms at the same time. The model introduced in (Kerr et al., 2000) is as follows:

$$log(y_{ijkg}) = \mu + A_i + D_j + V_k + G_g + (AG)_{ig} + (VG)_{kg} + \epsilon_{ijkg}. \tag{1.8}$$

Here, μ is the overall average signal, A_i represents the effect of the ith array, D_j represents the effect of the jth dye, V_k represents the effect of the kth variety, G_g represents the effect of the gth gene, $(AG)_{ig}$ represents a combination of array i and gene g, and $(VG)_{kg}$ represents the interaction between the kth variety and the gth gene. The error term ϵ_{ijkg} is assumed to be independent and identically distributed with mean zero. After estimating the coefficients, the gene-(sample)variety term VG_{kg} is the parameter we want to infer. Note that this approach also assumes linear normalization factors.

Some previous studies established probabilistic normal (Chen et al., 1997) or Gamma (Newton et al., 2000) model on the intensities and performed normalization and expression level assessment based on the model. Parametric approaches, however, may suffer from model fitting problems when data generated in different labs do not support the model.

Another possible systematic variation needed for consideration is gene-label interaction. For example, Cy3-dCTP may be preferentially incorporated into a specific sequence, relative to Cy5-dCTP. If such interaction exists, certain genes will always show higher intensity in one of the channels, even under nondifferential expression conditions and after normalization. In such a case, the normalized log-ratios in different slides of calibration experiment will be correlated and these correlations can be used to detect gene-label interactions. Table 1.3 demonstrates that, except for C1S1$\sim$2 in 125-gene project, the residuals are poorly correlated between different

Table 1.3 *High correlation indicates the phenomenon of gene-label interactions.*[a]

	125-gene project					4129-gene project			
	C1S1	C1S2	C3S1	C4S1	C4S2		C1S1	C1S2	C2S1	C2S2
C1S1	1.00	0.84	0.12	0.31	−0.13	C1S1	1.00	0.21	0.20	0.12
C1S2	0.84	1.00	0.07	0.37	−0.17	C1S2	0.21	1.00	0.17	0.13
C3S1	0.12	0.07	1.00	0.17	−0.21	C2S1	0.20	0.17	1.00	0.31
C4S1	0.31	0.37	0.17	1.00	0.36	C2S2	0.12	0.13	0.31	1.00
C4S2	−0.13	−0.17	−0.21	0.36	1.00	—	—	—	—	—

[a]Except for C1S1 and C1S2 in the 125-gene project, the gene-label interactions are not significant.

slides in both 125-gene and 4129-gene projects. Theoretically, some degree of gene-label interaction may exist; however, this interaction appears to be insignificant in magnitude compared with other sources of variation in the present experiment.

1.2.7 Hierarchical linear model

When we have hierarchical data as illustrated in Figure 1.16, a Bayesian approach of the hierarchical linear model can be used to incorporate prior knowledge generated from calibration experiments as well as to account for variations introduced at different experiment stages. In Bayesian analysis, the prior knowledge is first used to construct prior distributions of unobserved parameters. The posterior distribution of the desired parameters is then computed to represent the combined information on the parameters from the observed data and the specified prior distributions. Finally, we can perform analysis on the desired parameters based on the resulting posterior distribution.

1.2.8 Model formulation

Denote by x_{gse} the normalized log-ratios of gene g, slide s, calibration experiment e and y_{gse} the normalized log-ratios of gene g, slide s, comparative experiment e. We recognize that y_{gse} is affected by the slide-effect and uncontrollable variation between the different bacteria cultures used in different experiments. For each experiment (culture), y_{gse} is a sampling from a normal distribution of slide effect within the same culture. Thus, $y_{gse} \sim N(\mu_{ge}, {\tau_g}^2)$, where μ_{ge} is the mean among different slides within this culture, and ${\tau_g}^2$ is the variance of slide-effect distribution for gene g. Furthermore, the within-experiment mean μ_{ge} is, in turn, a sampling from a normal distribution of culture variation. Thus, $\mu_{ge} \sim N(\theta_g, {\sigma_g}^2)$ where θ_g measures the true log-fold-change of gene g and ${\sigma_g}^2$ is the variance between bacteria cultures. The probability model of the hierarchical structure is

$$y_{gse} \sim N(\mu_{ge}, {\tau_g}^2), \qquad \mu_{ge} \sim N(\theta_g, {\sigma_g}^2). \tag{1.9}$$

Note that only y_{gse} are observed data while ${\tau_g}^2$, ${\sigma_g}^2$, and θ_g are unobserved parameters. Under this model, θ_g is the unknown parameter of interest and the derived posterior distribution of θ_g will be used to assess the expression level of gene g. If gene g is nondifferentially expressed, then the posterior of θ_g is distributed around 0. Intuitively, to declare a gene differentially expressed means that the y_{gse} deviate from 0 in the same direction and that the deviations are large compared to the magnitude of the posterior distribution of ${\tau_g}^2$ and ${\sigma_g}^2$.

The information pooled across the calibration slides is then used to obtain a prior distribution for the slide-effect variance and culture effect variance:

$$ {\tau_g}^2 \sim \frac{k\tilde{\tau_g}^2}{{\chi_k}^2}, \qquad {\sigma_g}^2 \sim \frac{h\tilde{\sigma_g}^2}{{\chi_h}^2}, \qquad \theta \propto 1. \tag{1.10} $$

Here, $\tilde{\tau_g}^2 = ((S-1)E\hat{\tau_g}^2 + \hat{\tau_A}^2)/(S-1)E+1$ is the weighted value of gene-specific and overall sample variances obtained from calibration slides; $\hat{\tau_g}^2 = \sum_{s,e}(x_{gse} - x_{g\cdot e})^2/(S-1)E$ ($x_{g\cdot e} = mean_s(x_{gse})$), $\hat{\tau_A}^2 = \sum_{g,s,e}(x_{gse} - x_{g\cdot e})^2/\,G(S-1)E$, G the total number of genes, S the number of slides, E the total number of calibration experiments, and ${\chi_k}^2$ the chi-square distribution with degree of freedom k and k an adjustable degree of freedom, similarly, $\tilde{\sigma_g}^2 = (E\hat{\sigma_g}^2 + \hat{\sigma_A}^2)/(E+1)$, $\hat{\sigma_g}^2 = \sum_g {x_{g\cdot e}}^2/E$, $\hat{\sigma_A}^2 = \sum_{g,e} {x_{g\cdot e}}^2/GE$, ${\chi_h}^2$ the chi-square distribution with degree of freedom h and h an adjustable degree of freedom. We note that $\tilde{\sigma_g}^2$ is biased upward as an estimate of ${\sigma_g}^2$. As a result, our procedure will tend to be conservative.

Because the posterior distributions of the parameters do not have closed form solution, a Markov chain Monte Carlo method (MCMC) (Gilks et al., 1995) is used to simulate the desired posterior distributions. The full conditional distributions used in the MCMC simulation (known as Gibbs sampling) is as follows:

1. Initialize $(\mu_{ge})^{(0)} = y_{g\cdot e}$ and $(\theta_g)^{(0)} = (\mu_{g\cdot})^{(0)}$.
2. Generate $({\sigma_g}^2)^{(i)}$ from distribution ${\sigma_g}^2 \mid (\mu_{ge})^{(i-1)}$ where

$$ {\sigma_g}^2 \mid \mu_{ge} \sim \frac{\sum_e(\mu_{ge} - \mu_{g\cdot})^2 + h\tilde{\sigma_g}^2}{{\chi_{E+h-1}}^2}. $$

3. Generate $(\theta_g)^{(i)}$ from distribution $\theta_g \mid (\mu_{ge})^{(i-1)}, ({\sigma_g}^2)^{(i)}$ where

$$ \theta_g \mid \mu_{ge}, {\sigma_g}^2 \sim N\left(\mu_{g\cdot}, \frac{{\sigma_g}^2}{E}\right). \tag{1.11} $$

4. Generate $({\tau_g}^2)^{(i)}$ from distribution ${\tau_g}^2 \mid (\mu_{ge})^{(i-1)}, y_{gse}$ where

$$ {\tau_g}^2 \mid \mu_{ge}, y_{gse} \sim \frac{\sum_{e=1}^{E}\sum_{s=1}^{S_e}(y_{gse} - \mu_{ge})^2 + k\tilde{\tau_g}^2}{\chi^2_{S_1+\ldots+S_E+k}}. \tag{1.12} $$

   S_e is the number of slides in experiment e.

5. Generate $(\mu_{ge})^{(i)}$ from distribution $\mu_{ge} \mid y_{gse}, ({\tau_g}^2)^{(i)}, (\theta_g)^{(i)}, ({\sigma_g}^2)^{(i)}$ where

$$\mu_{ge} \mid y_{gse}, {\tau_g}^2, \theta_g, {\sigma_g}^2 \sim N\left(\frac{S_e y_{g\cdot e}{\sigma_g}^2 + {\tau_g}^2\theta_g}{S_e{\sigma_g}^2 + {\tau_g}^2}, \frac{{\tau_g}^2{\sigma_g}^2}{S_e{\sigma_g}^2 + {\tau_g}^2}\right). \tag{1.13}$$

6. Repeat procedures 2 through 5 for N times. We found that $N = 4000$ is sufficient for the mixing of the Markov Chain with a steady-state distribution that is the desired posterior distribution.

The preceding model can also be applied when calibration experiments are not available to provide prior information. In such cases, we assume the same hierarchical model with prior distribution ${\tau_g}^2 \sim k\tilde{\tau}^2/{\chi_k}^2$ and ${\sigma_g}^2 \sim h\tilde{\sigma}^2/{\chi_h}^2$ where $\tilde{\tau}^2 = \sum_{g,s,e}(y_{gse} - y_{g\cdot e})^2/G(S-1)E$ and $\tilde{\sigma}^2 = \sum_{g,e}(y_{g\cdot e} - y_{g\cdot\cdot})^2/G(E-1)$ become non-gene-specific. When S and E are small relative to prior degree of freedom h and k, the posterior will tend to be non-gene-specific while if S and E are large, the posterior distributions are dominated by gene-specific observations.

The hierarchical model has several assumptions including common slide variation in different experiments, a uniform prior on θ_g and normality. The normal assumption is validated by QQ-plot in (Tseng et al., 2001) and is generally expected to be a proper assumption. The model can be improved by applying experiment-specific slide variation, a reasonable center concentrated distribution on θ_g or a more sophisticated slide and experiment distribution. The improvement will, however, greatly slow down the MCMC simulation. Because the simulations are needed for every gene independently, it will require considerable time to analyze the whole genome. Currently, the simulation converges quickly in 4000 simulations in contrast to tens of thousands of simulations needed in other common hierarchical models due to the simple normal assumption and the conjugate prior. But even in this simpler case, it takes about 20 minutes to run in C and several hours to run in R for $\sim$4000 genes (Pentium III 866, 512MB RAM). Thus modifying the model may burden the computation to be impracticable.

Figure 1.21 depicts the 95% posterior interval of θ_g (i.e., intervals containing 95% of the probability in the posterior distribution of θ_g) on common genes in both the 125-gene project and the 4129-gene project. We note that genes with stronger agreement in normalized log-ratios across the two replicated comparative experiments have shorter intervals, as expected. The 4129-gene project has generally larger intervals than the 125-gene project, perhaps because the former is singly spotted and lacks a quality filtering step.

The results of two projects show general agreement. According to the 95% posterior interval, among 119 common genes of our two projects, there are 35 up-regulated, 30 down-regulated genes in the 125-gene project and 23 up-regulated, 19 down-regulated genes in the 4129-gene project. Among them, there are 17 up-regulated and 17 down-regulated genes that agree in both projects. The average length of 95% intervals of normalized log-ratios are 0.27 and 0.43, respectively, in the 125-gene and 4129-gene projects that correspond to 0.73- to 1.4- and 0.61- to 1.6-fold changes, respectively. In the few strong disagreement genes of two projects, we found that most

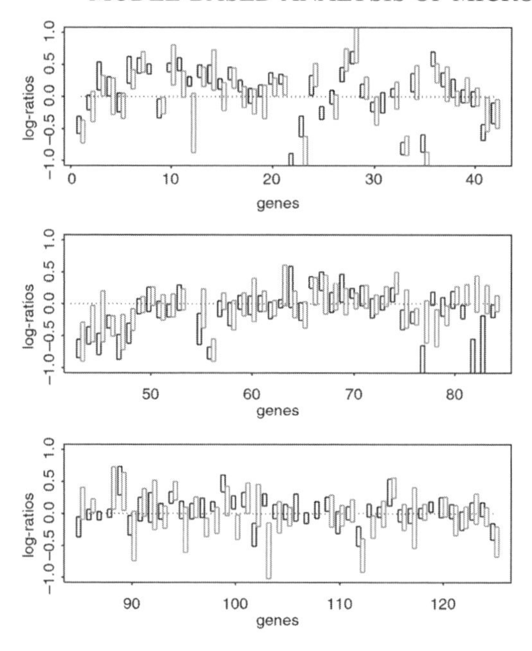

Figure 1.21 *(See color insert following page 114.) The orange and green rectangles show the 95% posterior interval for the underlying expression level* θ_g *of the 125-gene and 4129-gene projects (green: 125-gene project; orange: 4129-gene project). Rectangles of gene 54 (aceA) are below* -1.0 *and do not appear in the graph.*

of them are grouped in some pathways, such as metE, metB, aroL, aroG, and aroF. This suggests that these strong disagreements may reflect real biological variation between the cultures used in the two different projects. We have not discussed how to account for multiple comparisons (i.e., selecting apparently differentially expressed genes from the large number of genes in the genome). Methods to account for multiple comparisons are reviewed in Dudoit et al. (2002b) and in the next chapter.

1.2.9 Discussion of some experimental design issues

Reverse-labeling and calibration experiment

In a reverse-labeling design (Marton et al., 1998; Kerr et al., 2000), each of the two samples (for example, A and B) to be compared is divided into two aliquots and labeled with two different dyes (for example, Cy3 and Cy5) in separate steps. Two hybridization experiments are then performed. In the first hybridization solution, sample A is Cy3-labeled and sample B is Cy5-labeled. In the second hybridization solution, the labeling is reversed. We can use our calibration experiments to assess the usefulness of reverse-labeling by regarding the results of the two slides in a calibration experiment, say C1S1 and C1S2, as arising from the two hybridizations of a reverse-labeling experiment. This is valid because, in this case, all four labeling reactions were performed on aliquots derived from the same sample. Thus, the calibration experiment is a special case of reverse-labeling when the comparative samples A and B are identical. The first two plots of Figure 1.22 give the scatter plot of the difference (log(Cy5) − log(Cy3)) vs. the average [log(Cy3) + log(Cy5)]/2. The systematic trends that are evident in these plots are due to the inadequacy of linear normalization. As a result, if an ordinary design were used, then low-expression genes in the Cy5-labeled comparative sample are likely to be incorrectly identified as being down-regulated; however, this problem is greatly alleviated by the reverse-labeling design. The estimated gene effect (log(Cy5)–log(Cy3)) from these two slides (the third plot of Figure 1.22) cluster tightly around the zero line and show no systematic trend, just as it should be when the two comparative samples are identical. Thus, in this example, reverse-labeling offers useful protection against the nonlinearity of label normalization without the need to explicitly model the nonlinearity. The analysis in (Tseng et al., 2001) shows that such protection is not guaranteed, but partial protection can be expected under the condition that the nonlinearity contributions of each gene have the same sign in both slides. Another potential benefit of reverse-labeling is the cancellation of gene-label interaction. Gene-label interaction can also be

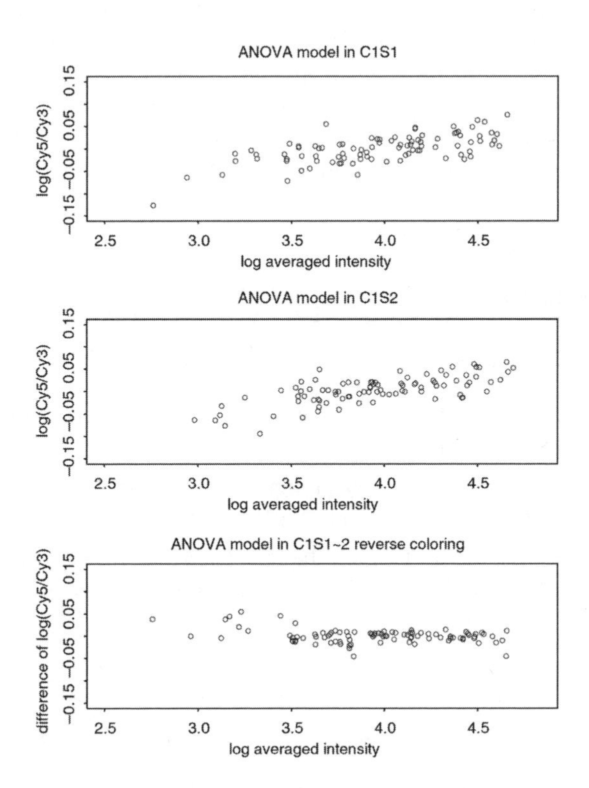

Figure 1.22 *An increasing trend occurs in both the first and second plot. When applying reverse labeling design, the trend is greatly canceled.*

handled through explicit modeling, but this has not been pursued because gene-label interaction compared to the other sources of variation is not significant in most data.

Note that the reverse-labeling design has the advantage of simple computation; however, when we want to perform a series of experiments such as taking samples at different time points, the design will be more cumbersome. Performing nonlinear normalization and explicit modeling of gene-label interaction is a useful alternative.

Multiple spotting vs. multiple slides

Multiple spots and multiple slides are replications to help us assess variations due to spots and slides. Normalization is slide-dependent, therefore, multiple slide information cannot be used to assess experiment quality before normalization. Thus, the normalization procedure itself is vulnerable to contamination by poor quality spots. On the other hand, multiple spots within the same slide provide useful information for filtering out contaminated spots, poor quality genes, or problematic slides in each experiment. We also tried to apply a similar quality filtering procedure to normalized log-ratios in singly spotted replicate arrays. This is less effective because the between slide variation is typically much larger than between spot variation, thus reducing the power for detection of outliers. In practical microarray applications, it may be desirable to monitor as many genes as possible at the beginning and singly spotted arrays are more effective at this stage. After narrowing down the number of target genes, however, one may be interested in using a custom array to investigate these genes further. The use of multiple spotting should be considered in the design of these arrays.

1.3 Acknowledgments

We thank the following for permission to reproduce their work:

- Figure 1.2 through Figure 1.10 — are from Li, C. and W.H. Wong, *Proc. Natl. Acad. Sci.*, 98: 31–36, 2001a.
- Figure 1.11 through Figure 1.15 and Table 1.1 are from Li, C. and W.H. Wong, *Genome Biol.*, 2(8):0032.1–0032.11, 2001b.
- Figure 1.17 through Figure 1.22 and Table 1.2 and Table 1.3 are from Tseng, G.C. et al., *Nucleic Acids Res.*, 29(12):2549–2557, 2001.

CHAPTER 2

Design and analysis of comparative microarray experiments

Yee Hwa Yang and Terry Speed

2.1 Introduction

This chapter discusses the design and analysis of relatively simple comparative experiments involving microarrays. Some of the discussion applies to all the most widely used kinds of microarrays, that is, radiolabeled cDNA arrays on nylon membranes, two-color, fluorescently labeled cDNA or long oligonucleotide arrays on glass slides, or single color, fluorescently labeled, high-density short oligonucleotide arrays on silicon chips. The main focus, however, is on two-color complementary deoxyribonucleic acid (cDNA) or long oligonucleotide arrays on glass slides because they present more challenging design and analysis problems than the other two kinds.

As subfields of statistics, the topics of design and analysis of microarray experiments are still in their infancy. Entirely satisfactory solutions to many simple problems still elude us, and the more complex problems will provide challenges to us for some time to come. Much of what we present in this chapter could be described as first pass attempts to deal with the deluge of data arriving at our doors. Questions come in a volume and at a pace that demands answers; we simply do not have the luxury of waiting until we have final solutions to problems before we get back to the biologists. A major aim of this chapter is to stimulate other statisticians to work with their local biologists on microarray experiments and to come up with better solutions to the common problems than the ones we present here.

For software related to the topics discussed in this chapter, we refer to Parmigiani et al. (2003) and the website `http://astor.som.jhmi.edu/nox/pgiz.html`.

2.2 Experimental design

Statisticians do not need reminding that proper statistical design is essential to ensure that the effects of interest to biologists in microarray experiments are accurately and precisely measured. Much of our approach to the design (and analysis, see Section 2.3)

1-58488-327-8/03/$0.00+$1.50
© 2003 by CRC Press LLC

of microarray experiments, takes as its starting point the idea that we are going to measure and compare the expression levels of a single gene in two or more cell populations. The fact that, with microarrays, we do this simultaneously for tens of thousands of genes definitely has implications for the design of these experiments, but initially we focus on a single gene.

In this section, some of the design issues that arise with the two-color cDNA or long oligonucleotide microarray experiments are discussed. Designing experiments with radiolabeled cDNA arrays on nylon membranes or for fluorescently labeled, high-density short oligonucleotide arrays is less novel. Apart from the following brief discussion about probe design, and a few other remarks in passing, we will not discuss these two platforms in detail separately.

Any microarray experiment involves two main design aspects: the design of the array itself, that is, deciding which DNA probes are to be printed on the solid substrate, be it a membrane, glass slide or silicon chip, and where they are to be printed; and the allocation of messenger ribonucleic acid (mRNA) samples to the microarrays, that is, deciding how mRNA samples should be prepared for the hybridizations, how they should be labeled, and the nature and number of the replicates to be done. We focus on the second aspect, after making a few remarks about the first.

The choice of which DNA probes to print onto the solid substrate is usually made prior to consulting a statistician; this choice is determined by the genes with expression levels that the biologist wants to measure, or by the cDNA libraries (that is, the collections of cDNA clones) available to them. With high-density short oligonucleotide arrays, these decisions are generally made by the company (e.g., Affymetrix) producing the chips, although opportunities exist for building customized arrays. Many researchers purchase pre-spotted cDNA slides or membranes in the same way as they do high-density short oligo arrays. With short (25 base pair) or long (60–75 base pair) oligonucleotide microarrays, the determination of the probe sequences to be printed is an important and specialized bioinformatic task (see Hughes et al. (2001); Rouillard et al. (2001); `http://www.affymetrix.com/technology/design/index.affx` for a discussion). Similarly, many issues need to be taken into account with cDNA libraries of probes, and here we refer to Kawai et al. (2001).

Advice is sometimes sought from statisticians on the use of controls: negative controls such as blank spots, spots with cDNA from very different species (e.g., bacteria when the main spots are mammalian cDNA), or spots "printed" from buffer solution, or positive controls such as so-called "housekeeping" genes that are ubiquitously expressed at more or less constant levels, and genes that are known not to be in the target samples, which are to be spiked into it. We note that commercially produced chips (e.g., by Affymetrix) have a wide range of controls of these kinds built in. The questions typically posed to statisticians concern the nature and number of such controls, and the use to be made of signal from them in later analysis. Some controls are there to reassure the experimenter that the hybridization was a success, or indicate that it was a failure, as the case may be. Others are to facilitate special tasks such as normalization (see Chapter 1) or to permit an assessment of the quality of the experimental results. Controls used with cDNA microarrays for normalization include the so-called microarray sample pool (see Yang et al. (2002b) and the ScoreCard system from Amersham Biosciences). At this stage we do not have enough experience to offer general advice or conclusions in design for controls as the full use of such control data in cDNA arrays is still in its early stages. Yet another cDNA spot design issue on which statisticians might be consulted is the replicating of spots on the slide, and we discuss this in Section 2.2.4.

2.2.1 Graphical representation

First, we introduce a method for graphical representation of microarray experimental designs. One convenient way to represent microarray experiments is to use a *multi-digraph*, which is a directed graph with multiple edges as illustrated in Figure 2.1(a). In such a representation, vertices or nodes (e.g., A, B) correspond to target mRNA samples and edges or arrows correspond to hybridizations between two mRNA samples. By convention, we place the green-labeled sample at the tail and the red-labeled sample at the head of the arrow. For example, Figure 2.1(a) depicts an experiment consisting of replicated hybridizations. Each slide involves labeling sample A with green (e.g., Cy3) dye, sample B with red (e.g., Cy5) dye and hybridizing them together on the same slide. The number "5" on the arrow indicates the number of replicated hybridizations in this experiment. Similar graphical representations of this nature have been used previously in experimental design, for instance, in the context of measurement agreement comparisons (Youden, 1969). For the rest of this chapter, we use this representation to illustrate different microarray designs. The structure of the graph determines which effects can be estimated and the precision of the estimates. For example, two target samples can be compared only if an undirected path joins the corresponding two vertices. The precision of the estimated contrast then depends on the number of paths joining the two vertices and is inversely related to the length of the path. In the hypothetical experiment presented in Figure 2.1(b), which consists of three sets of hybridizations, the number of paths joining the vertices A and B is

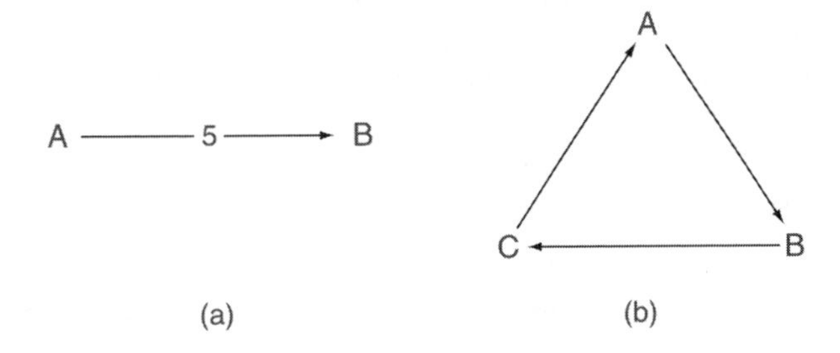

Figure 2.1 *Graphical representation of designs. In this representation, vertices correspond to target mRNA samples and edges to hybridizations between two samples. By convention, we place the green-labeled sample at the tail and the red-labeled sample at the head of the arrow. The number 5 denotes the number of replicates of that hybridization.*

2; a path of length 1 runs directly between A and B; another path of length 2 joins A and B via C. When we are estimating the relative abundance of target samples A and B, the estimate of $\log_2(A/B)$ from the path A to B is likely to be more precise than the estimate of $\log_2(A/B)$ by $\log_2(A/C) - \log_2(B/C)$ from the path of length 2 joining A and B via C.

The preceding discussion assumes that the spot intensities in two-color experiments are all reduced to ratios before further analysis; however, it is already the case that some authors (Jin et al., 2001) are using single-channel data and not reducing to ratios. In this case, two strata of information exist on the log scale: the usual log-ratios within hybridizations, and log-ratios between hybridizations. Because of the novelty of this analysis approach, and the absence so far of a thorough discussion of single channel normalization, we concentrate our discussion of design and analysis to what in effect, is the within-hybridization stratum (i.e., to analyses that depend on log-ratios within slides).

2.2.2 *Optimal designs*

All measurements from two-color microarrays are paired comparisons, that is, measurements of relative gene expression, with a microscope slide playing the role of the block of two units. We begin by discussing design choices for simple experiments comparing two samples T and C. These include experiments comparing treated and untreated cells (e.g., drug treated and controls), cells from mutant (including knock-out or transgenic), and from *wild-type* organisms (Callow et al., 2000), or cells from two different tissues (e.g., tumor and normal). Suppose that we wish to compare the expression level of a single gene in the samples T and C of cells. We could compare them on the same slide (i.e., in the *same* hybridization) in which case a measure of the gene's differential expression could be $\log_2(T/C)$, where $\log_2 T$ and $\log_2 C$ are measures of the gene's expression in samples T and C. We refer to this as a *direct* estimate of differential expression — direct because the measurements come from the same hybridization. Alternatively, $\log_2 T$ and $\log_2 C$ may be estimated in two *different* hybridizations, with T being measured together with a third sample R and C together with another sample R' of R, on two different slides. The log-ratio $\log_2(T/C)$ will in this case be replaced by the difference $\log_2(T/R) - \log_2(C/R')$, and we call this an *indirect* estimate of the gene's differential expression because it is calculated with values $\log_2 T$ and $\log_2 C$ from different hybridizations. Figure 2.2 presents the graphical representation of these two designs.

The early microarray studies (DeRisi et al., 1996; Spellman et al., 1998; Perou et al., 1999; and many others) performed their experiments using *indirect* designs. These designs are also known as *common reference designs* in the microarray literature because each mRNA sample of interest is hybridized together with a common reference sample on the same slide. The common reference samples could be tissues from *wild-type* organisms, or *control* tissue, or it could be a *pool* of all the samples of interest. Common references are frequently used to provide easy means of comparing

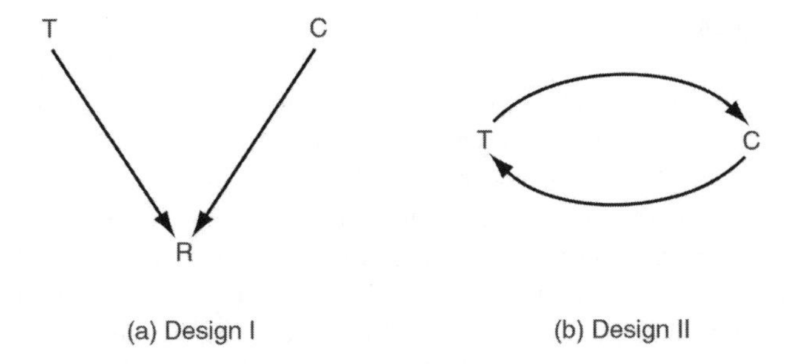

Figure 2.2 *Two possible designs comparing the gene expression in two samples T and C of cells. (a) Indirect comparison: This design measures the expression levels of samples T and C separately on two different slides (hybridizations) and estimates the log-ratio $\log_2(T/C)$ by the difference $\log_2(T/R) - \log_2(C/R)$. (b) Direct comparison: This design measures the gene's differential expression in samples T and C directly on the same slide (hybridization).*

many samples against one another. More recently, several studies (Jin et al., 2001; Kerr et al., 2001; Lin et al., 2002) have performed experiments that provide *direct* estimates of log-ratios. In such cases, fixed or random effect linear models and analysis of variance (ANOVA) have been used to combine data from the different hybridizations.

To date, the main work on design of two-color microarray experiments is due to Kerr and Churchill (2001), and Glonek and Solomon (2002), who have applied ideas from optimal experimental design to suggest efficient designs for some of the common cDNA microarray experiments. Kerr and Churchill (2001) based their comparisons of different designs on the A-optimality criterion. In addition, they introduced a novel class of designs they called *loop* designs, and found that under A-optimality, *loop* designs were more efficient than common reference designs.

Suppose we have a single factor experiment with K levels and the goal is to compare all K treatments. The A-optimality criterion favors designs that minimize the average variance of contrasts of interest; however, this criterion alone is often not enough to single out one design; see Designs V and VI in table of Yang and Speed (2002). Just as different microarray experiments will require different analyses, no best design class suits all experiments. Frequently, the scientific questions and physical constraints will drive the design choices.

Glonek and Solomon (2002) studied optimal designs for time course and factorial experiments. Their article introduced classes of appropriate designs based on the notion of *admissibility*. For the same number of hybridizations, a design is said to be *admissible* if there exists no other design that has a smaller variance for all contrasts of interest. Their idea is that an investigator should compare only the admissible designs and then base their design selection on scientific interest. In Glonek and Solomon (2002) and in other similar calculations in the literature, log-ratios from different experiments are regarded as statistically independent. In Section 2.3.9, we revisit these calculations assuming a more realistic covariance between replicates, and we examine the implications for design optimality.

2.2.3 Design choices

In preparing to design a cDNA microarray experiment, certain general issues need to be addressed. These can be separated into scientific and logistic (practical). The scientific issues include the aim of the experiment. It is most important to state the primary focus of the experiment, which may be identifying differentially expressed genes, searching for specific patterns, or identifying tumor subclasses. Results from previous experiments or other prior knowledge may lead us to expect only a few, or many genes differentially expressed. In addition, there may be multiple aims within a single experiment, and it is important to specify the different questions and priorities between them. Practical or logistic issues include information such as the types of the mRNA samples, the amount of material and the number of slides (chips) available. The source of mRNA (e.g., tissue samples or cell lines) will affect the amount of mRNA available, and in turn the number of replicate slides possible.

Other information to keep in mind includes the experimental process prior to hybridization such as sample isolation, mRNA extraction, amplification and labeling. These and other technical matters are discussed in Schena (2000) and Bowtell and Sambrook (2002). Keeping track of all the different aspects of the experimental process helps us better understand the different levels of variability affecting our microarray data. Finally, consideration should be given to the verification method following the experiments, such as Northern or Western blot analysis, real-time polymerase chain reaction (PCR), or *in-situ* hybridization. The amount of verification to be carried out can influence statistical methods used and the determination of sample size. All this information helps us translate an experiment's biological goals into the corresponding statistical questions and then, following appropriate design choices, helps us obtain a ready interpretation of the results.

We begin our discussion of the design of experiments when there is just one natural design choice, when one design stands out as preferable to all others, given the nature of the experiment and the material available. For example, suppose that we wish to study mRNA from two or more populations of cells, each treated by a different drug, and that the primary comparisons of interest are those of the treated cells versus the untreated cells. In this case, the appropriate design is clear: the untreated cells become a de facto reference, and all hybridizations involve one treated set of cells and the untreated cells. Next, suppose that we have collected a large number of tumor samples from patients. If the scientific focus of the experiment is on discovering tumor subtypes (Alizadeh et al., 2000), then the design involving comparisons between all the different tumor samples and a common reference RNA is a natural choice. In both cases, the choice follows from the aim of the study, with statistical efficiency considerations playing only a small role.

The statistical principles of experimental design are randomization, replication, and local control; naturally, these all apply to two-color microarray experiments, especially the last two. We have so far found only limited opportunities for randomization, however, the development of appropriate ways of randomizing microarray experiments would be a useful research project. In this and many similar laboratory contexts, the challenge is to balance the requirements of uniformity (e.g., of reagents, techniques, technicians, perhaps even time of day for the experiment), which aim to reduce unnecessary variation, with the statistical need to provide valid estimates of experimental error. The situation recalls the discussion between R.A. Fisher and W.S. Gosset ("Student") in the 1930s concerning the relative merits of random and systematic layouts for field experiments; see Pearson and Wishart (1958) and Bennett (1971–1974) for the papers. A number of issues are highly specific to experimental design in the microarray context, and we now turn to a brief discussion of some of them. Parts of what follows will also be relevant to nylon membrane and high-density oligonucleotide arrays.

2.2.4 Replication

As indicated earlier, and consistent with statistical tradition (Fisher, 1926), replication is a key aspect of comparative experimentation, its purpose being to increase precision and, more important, to provide a foundation for formal statistical inference. In the microarray context, a number of different forms of replication occur. The differences are all in the degree to which the replicate data may be regarded as independent, and in the populations that experimental samples are seen to represent. Given that replicate hybridizations are almost invariably carried out by the same person, using the same equipment and protocols, and frequently at about the same time, it is inevitable that replicate data will share many features. Most of the differences discussed next concern the target mRNA samples.

Duplicate spots

Many groups print cDNA in duplicate on every slide, frequently in adjacent positions. At times, even greater within-slide replication is used, particularly with smaller customized rather than the larger general clone sets. This practice provides valuable quality information, as the degree of concordance between duplicate spot intensities or relative intensities is an excellent quality indicator; however, because replicate spots on the same slide, particularly adjacent spots, will share most if not all their experimental conditions, the data from the pairs cannot be regarded as independent. Although averaging log-ratios from duplicate spots is appropriate, their close association means that the information is less than that from pairs of truly independent measurements. Typically, the overall degree of concordance between duplicate spots is noticeably greater than that observed between the same spot across replicate slides, although exceptions exist.

Technical replicates

This term is used to denote replicate hybridizations where the target mRNA is from the same pool (i.e., from the same biological extraction). It has been observed that characteristic, repeatable features of extractions exist, and this leads us to conclude that technical replicates generally involve a smaller degree of variation in measurements than the biological replicates described next. Usually, the term technical replicate includes the assumption that the mRNA sample is labeled independently for each hybridization. A more extreme form of technical replication would be when a sample from the one extraction *and* labeling is split, and replicate hybridizations done with subsamples of this kind. We do not know of many labs now doing this, though some did initially. Section 2.3.9 discusses in more detail how technical replicates affect design decisions.

Biological replicates — type I

This term refers to hybridizations involving mRNA from different extractions, for example, from different samples of cells from a particular cell line or from the same tissue. In most cases, this will be the most convenient form of genuine replication.

Biological replicates — type II

This term is used to denote replicate slides where the target mRNA comes from the same tissue but from different individuals in the same species or inbred strain, or from different versions of a cell line. This form of biological replication is different in nature from the type I biological replicates described previously, and typically involves a much greater degree of variation in measurements. For example, experiments with inbred strains of mice have to deal with the inevitability of different mice having their hormonal and immune systems in different states, the tissues exhibiting different degrees of inflammatory activity, and so on. With noninbred individuals, the variation will be greater still.

The type of replication to be used in a given experiment depends on the *precision* and on the *generalizability* of the experimental results sought by the experimenter. In general, an experimenter will want to use biological replicates to support the generalization of his or her conclusions, and perhaps technical replicates to reduce the variability in these conclusions. Given that several possible forms of technical and biological replication usually exist, judgment will need to be exercised on the question of how much replication of a given kind is desirable, subject to experimental and cost constraints. For example, if a conclusion applicable to all mice of a certain inbred strain is sought, experiments involving multiple mice, preferably a random sample of such mice, must be performed.

Note that we do not discuss sample size determination or power in this chapter. Despite the existence of research showing how to determine sample size for microarray experiments based on power considerations, we do not believe that this is possible. Our reasons are outlined in Yang and Speed (2002).

2.2.5 Dye-swaps

Most two-color microarray experiments suffer from systematic differences in the red and green intensities which require correction. Details of normalization are discussed in Chapter 1. In practice, it is very unlikely that normalization can be done perfectly for every spot on every slide, leaving no residual color bias. To the extent that this occurs, not using dye-swap pairs will leave an experiment prone to a systematic color bias of an unknown extent. When possible, we recommend using dye-swap pairs. Alternatively, random dye assignments may be used, in effect including the bias in random error. A theoretical analysis of the practice of dye-swapping has yet to be presented.

2.2.6 Extensibility

Often, experimenters want to compare essentially arbitrarily many sources of mRNA over long periods of time. One method is to use a common reference design for all experiments, with the common reference being a "universal" reference RNA that is derived from a combination of cell lines and tissues. Some companies provide universal reference mRNA (see e.g., `http://www.stratagene.com/gc/universal_mouse_reference_rna.html`) while many individual labs create their own common reference pool. Common references provide extensibility of the series of experiments within and between laboratories. When an experimenter is forced to turn to a new reference source, it may be difficult to compare new experiments with previous ones that were performed based on a different reference source. The ideal common reference, therefore, is widely accessible, available in unlimited amounts, and provides a signal over a wide range of genes. In practice, these goals can be difficult to achieve. When a universal reference RNA is no longer available, it is necessary to carry out additional hybridizations, conducting what we term a linking experiment to connect otherwise unrelated data. More generally, linking experiments allow experimenters to connect previously unrelated experiments, with the number of additional hybridizations depending on the precision of conclusions desired. Suppose that in one series of experiments we used reference R_1, and that in another series we used reference R_2. The linking experiment compares R_1 and R_2, thus permitting comparisons between two sources, one of which has been co-hybridized with R_1 and the other with R_2; however, this ability comes at a price. Log-ratios for source A co-hybridized with reference R_1 can be compared to ones from source Z co-hybridized with reference R_2, only by combining $A \rightarrow R_1$ and $Z \rightarrow R_2$ together with $R_1 \rightarrow R_2$ through the identity $\log_2(A/Z) = \log_2(A/R_1) + \log_2(R_1/R_2) - \log_2(Z/R_2)$. In other words, cross-reference comparisons involve combining three log-ratios, with corresponding loss of precision, as the variance of $\log_2(A/Z)$ here is three times that of the individual log-ratios. Nevertheless, there will be times when cross-referencing is worthwhile, particularly when one notices that the variance of the linking term $\log_2(R_1/R_2)$ can be reduced to an extent thought desirable simply by replicating that experiment. The linking term in the identity would be replaced by the average of all such terms across replicates.

2.2.7 *Robustness*

Loosely speaking, we call a design robust if the efficiency with which effects of interest are estimated does not change much when small changes are made to the design, such as those following from the loss of a small number of observations, or a change in the correlation structure of the observations. It is not uncommon for hybridization reactions to fail in microarray experiments; in a context where mRNA is hard to obtain, an experiment may well have to proceed without repeating failed hybridizations. In such cases, robustness is a highly desirable property of the design. A situation to be avoided is one where key comparisons of interest are estimable only when a particular hybridization is successful. It follows that heavy reliance on direct comparisons is not so desirable. Preferable is the situation where all quantities of interest are estimated by a mix of direct and indirect comparisons, that is, where many different paths connect samples in the design graph.

A nonstandard way to improve the robustness of a design is to give careful thought to the order in which the different hybridizations in the experiment are carried out. More critical hybridizations could be done earlier, and full sets of hybridizations completed before further replicates are run, leaving the greatest opportunity for revising the design in the case of failed hybridizations. Note that this practice is contrary to the generally desirable practice of randomizing the order in which the parts of an experiment (here hybridizations) are carried out. In this context, such randomization is frequently achievable, but it is not popular with experimenters because it will often require a greater number of preparatory steps and so an increased risk of failure.

As we will see later, the use of technical replicates introduces correlation between measured intensities and relative intensities. The precise extent of this correlation is typically difficult to measure. A design where the performance is more stable across varying technical replicate correlations would usually be preferred to one that is more efficient for one range of values of the unknown correlations, but less efficient for another range of values. This is a different form of robustness.

2.2.8 *Pooling*

An issue arising frequently in microarray experiments is the pooling of mRNA from different samples. At times pooling is necessary to obtain sufficient mRNA to carry out a single hybridization. At other times, biologists wonder whether pooling improves precision even when it is not necessary. Is this a good idea?

To sharpen the question, suppose that we wish to compare mRNA from source A with that from source B, using three hybridizations. We could carry out three separate extractions and labelings from each of the sources, arbitrarily pair A and B samples, and do three competitive hybridizations, each a single A sample versus a single B sample. We would then average the results; see Section 2.3 for further discussion. Alternatively, we could pool the labeled mRNA samples, one pool for the A and one for the B samples, then subdivide each pool into three technical replicates, and carry out three replicate competitive hybridizations of pooled A versus pooled B. Again, the results of the 3 comparisons would be averaged. Which is better? An analogous question can be posed with the single-color hybridizations (high-density short-oligonucleotide arrays, and nylon membranes). An *a priori* argument can be made for either approach. Pooling may well improve precision, that is, reduce the variance of comparisons of interest. But does it do so at the price of permitting one sample (or a few) to dominate the outcome, and so give misleading conclusions overall? These are hard questions to answer.

We know of no experiment with two-color microarrays aimed at answering these questions, but we have seen the results of such an experiment with the Affymetrix technology (Han et al., 2002). There we saw that averaging pooled samples and then comparing across the types described previously was slightly more precise than doing the same thing with averaged results from single samples, and in that case there seemed to be no obvious biases from individual samples. Our conclusion was that the gain in precision arising from pooling probably does not justify the risks, and that it is probably better to be able to see the between-sample variation, rather than lose the ability to do so. It remains to be seen whether this conclusion will stand up over time, and whether it applies to two-color microarrays.

2.2.9 Our design focus

With most experiments, a number of designs can be devised that appear suitable for use, and we need some principles for choosing one from the set of possibilities. The remainder of this chapter focuses on the question of identifying differentially expressed genes, and discusses design in this context. The identification of differentially expressed genes is a question that arises in a broad range of microarray experiments (Callow et al., 2000; Friddle et al., 2000; Galitski et al., 1999; Golub et al., 1999; Spellman et al., 1998). The types of experiments include: *single-factor* cDNA microarray experiments, in which one compares transcript abundance (i.e., expression levels) in two or more types of mRNA samples hybridized to different slides. Time-course experiments, in which transcript abundance is monitored over time for processes such as the cell cycle, can be viewed as a special type of single-factor experiment with time being the sole factor. We discuss them briefly from this perspective. Factorial experiments, where two or more factors are varied across the mRNA are also of interest, and we discuss their design and analysis as well.

2.3 Two-sample comparisons

The simplest type of microarray experiment is the two sample or binary comparison, where we seek to identify genes that are differentially expressed between two sources of RNA. Such comparisons might be between knock-out and wild-type, tumor and normal, or treated and control cells. With "single color" systems, such as the nylon membranes and high-density short oligonucleotide arrays, the comparisons can be between results from two arrays, or two sets of arrays. With the two-color cDNA or long oligonucleotide arrays, the comparison can be within a single slide, across each of a set of replicate slides involving direct comparisons, or involving indirect comparisons between slides.

2.3.1 Case studies I and II

We illustrate the ideas of this section with two sets of two sample comparisons. These studies both aim to identify differentially expression between a mutant and a wild-type organism, but they do it differently. Both studies are with two-color cDNA microarrays. Case study I involves replicates of direct comparisons made within a slide. By contrast, case study II involves indirect comparisons between samples co-hybridized to a common reference mRNA.

In order to identify and remove systematic sources of variation in the measured expression levels and allow between-slide comparisons, the data for case studies I and II (as well as case studies III and IV introduced next) were normalized using the within-slide spatial and intensity dependent normalization methods described in Yang et al. (2002b). Normalization methods are discussed in more detail in Chapter 1, and we make no further mention of the topic, apart from noting that it is a critical preprocessing step with almost any microarray experiment.

Case study I: swirl zebrafish experiment

The results from the swirl zebrafish experiment were given to us by Katrin Wuennenberg-Stapleton from the Ngai Lab at University of California, Berkeley, while the swirl embryos themselves were generously provided by David Kimelman and David Raible from the University of Washington in Seattle. The experiment was carried out using zebrafish to study early development in vertebrates. Swirl is a point mutation in the BMP2 gene that causes defects in the organization of the developing embryo along its dorsal-ventral body axis. This results in a reduction of cells showing ventral cell fates (i.e., cell types that are normally formed only within the ventral aspect of the embryo), such as blood cells are reduced, whereas dorsal structures such as somites and notochord are expanded. A goal of this swirl experiment was to identify genes with altered expression in the swirl mutant compared to the wild-type zebrafish. The data are from four replicate slides: two sets of dye-swap pairs. For each of these slides, target cDNA from the swirl mutant was labeled using one of the Cy3 or Cy5 dyes and the target cDNA wild-type mutant was labeled using the other dye. Figure 2.3 is the graphical representation of this experiment.

In this case study, target cDNA was hybridized to microarrays containing 8848 cDNA probes. The microarrays were printed using 4×4 print-tips and are thus partitioned into a 4×4 matrix of sub-arrays. Each sub-array consists of a 22×24 spot matrix that was printed with a single print-tip. In this and the other studies discussed next, we call the spotted cDNA sequences "genes," whether or not they are actual genes, ESTs (expressed sequence tags), or cDNA sequences from other sources.

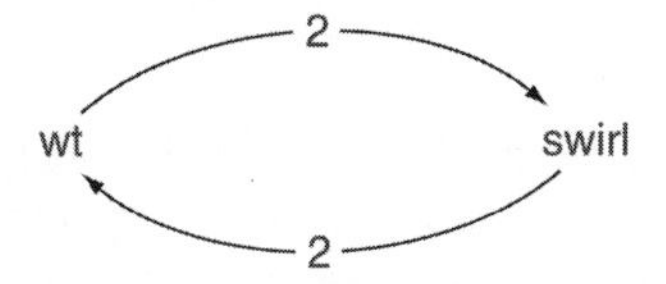

Figure 2.3 *Case study I: The swirl experiments provided by Katrin Wuennenberg-Stapleton from the Ngai Lab at the University of California, Berkeley. This experiment consists of two sets of dye swap experiments comparing gene expression between the mutant swirl and wild-type (wt) zebrafish. The number on the arrow represents the number of replicated experiments.*

Case study II: scavenger receptor BI mouse experiment

The scavenger receptor BI (SR-BI) experiment was carried out as part of a study of lipid metabolism and atherosclerosis susceptibility in mice, Callow et al. (2000). The SR-BI gene is known to play a pivotal role in high-density lipoprotein (HDL) metabolism. Transgenic mice with the SR-BI gene overexpressed have very low HDL cholesterol levels, and the goal of the experiment was to identify genes with altered expression in the livers of these mutant mice compared to "normal" FVB mice. The treatment group consisted of eight SR-BI transgenic mice, and the control group consisted of eight normal FVB mice. For each of these 16 mice, target cDNA was obtained from mRNA by reverse transcription and labeled using the red-fluorescent dye Cy5. The reference sample used in all 16 hybridizations was prepared by pooling cDNA from the eight control mice and was labeled with the green-fluorescent dye Cy3. The design would have been better if the reference sample had come from a *different* set of control mice. In this experiment, target cDNA was hybridized to microarrays containing 5548 cDNA probes, including 200 related to lipid metabolism. These microarrays were printed in a 4×4 matrix of sub-arrays, with each sub-array consisting of a 19×21 array of spots. Figure 2.4 is the graphical representation of this experiment. The data are available from `http://www.stat.Berkeley.EDU/users/terry/zarray/Html/srb1data.html`.

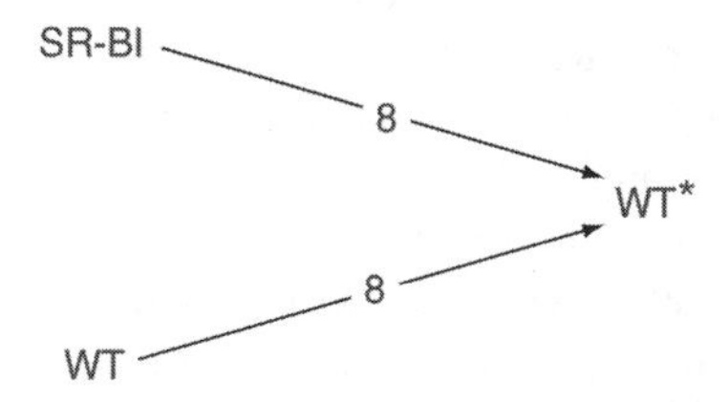

Figure 2.4 *Case study II: The SR–BI experiments provided by Matt Callow from the Lawrence Berkeley National Laboratory. This experiment consists of eight slides comparing gene expression between the transgenic SR–BI mice and the pooled control (WT*). Another eight slides comparing gene expression between normal FVB (WT) mouse and pooled control. The number on the arrow represents the number of replicated experiments.*

2.3.2 *Single-slide methods*

A number of methods have been suggested for the identification of differentially expressed genes in single-slide, two-color microarray experiments. In such experiments, the data for each gene (spot) consist of two fluorescence intensity measurements, (R, G), representing the expression level of the gene in the red (Cy5) and green (Cy3) labeled mRNA samples, respectively. (The most commonly used dyes are the cyanine dyes, Cy3 and Cy5, however, other dyes such as fluorescein and X-rhodamine may be used as well). We distinguish two main types of single-slide methods: those which are based solely on the value of the expression ratio R/G and those that also take into account overall transcript abundance measured by the product RG.

Early analyses of microarray data (DeRisi et al., 1996; Schena et al., 1995, 1996) relied on fold increase/decrease cutoffs to identify differentially expressed genes. For example, in their study of gene expression in the model plant *Arabidopsis thaliana*, Schena et al. (1995) use spiked controls in the mRNA samples to normalize the signals for the two fluorescent dyes (fluorescein and lissamine) and declare a gene differentially expressed if its expression level differs by more than a factor of 5 in the two mRNA samples. DeRisi et al. (1996) identify differentially expressed genes using a ± 3 standard deviation cutoff for the log-ratios of the fluorescence intensities, standardized with respect to the mean and standard deviation of the log ratios for a panel of 90 "housekeeping" genes (i.e., genes believed not to be differentially expressed between the two cell types of interest).

More recent methods have been based on statistical modeling of the (R, G) pairs and differ mainly in the distributional assumptions they make for (R, G) in order to derive a rule for deciding whether a particular gene is differentially expressed. Chen et al. (1997) propose a data dependent rule for choosing cutoffs for the red and green intensity ratio R/G. The rule is based on a number of distributional assumptions for the intensities (R, G), including normality and constant coefficient of variation. Sapir and Churchill (2000) suggest identifying differentially expressed genes using posterior probabilities of change under a mixture model for the log expression ratio $\log R/G$ (after a type of background correction, the orthogonal residuals from the robust regression of $\log R$ versus $\log G$ are essentially normalized log expression ratios). A limitation of these two methods is that they both ignore the information contained in the product RG. Recognizing this problem, Newton et al. (2001) consider a hierarchical model (Gamma–Gamma–Bernoulli model) for (R, G) and suggest identifying differentially expressed genes based on the posterior odds of change under this hierarchical model. The odds are functions of $R + G$ and RG and thus produce a rule which takes into account overall transcript abundance. The approach of Hughes et al. (2000b) (supplement) is based on assuming that R and G are approximately independently and normally distributed, with variance depending on the mean. It thus also produces a rule which takes into account overall transcript abundance.

As a result, each of these methods produces a model dependent rule which amounts to drawing two curves in the $(\log R, \log G)$-plane and calling a gene differentially expressed if its $(\log R, \log G)$-falls outside the region between the two curves. We

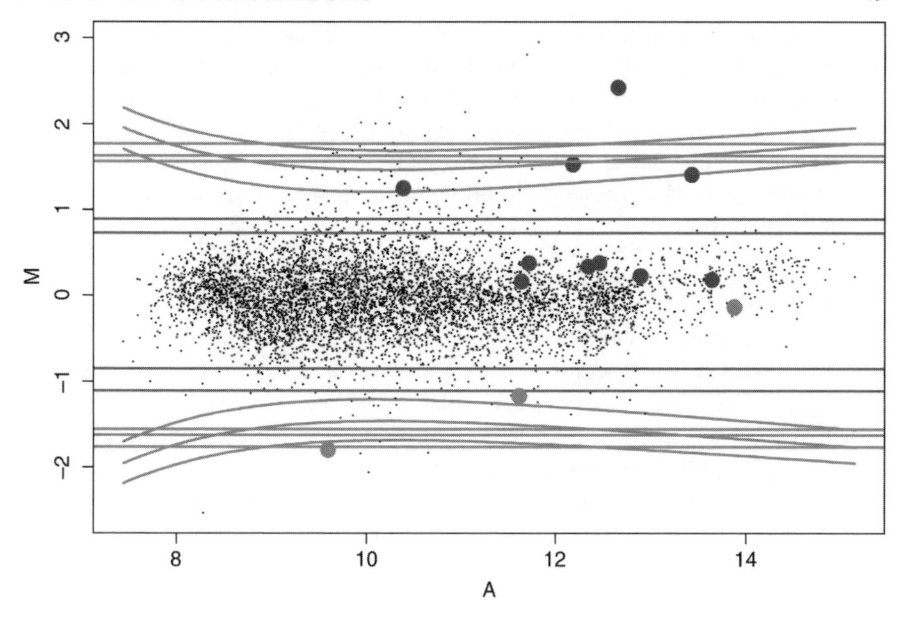

Figure 2.5 *(See color insert following page 114.) Single-slide methods: An MA–plot showing the contours for the methods of Newton et al. (2001) (orange, odds of change of 1:1, 10:1, and 100:1), Chen et al. (1997) (purple, 95% and 99% "confidence"), and Sapir and Churchill (2000) (cyan, 90%, 95%, and 99% posterior probability of differential expression). The points corresponding to genes with adjusted p-value less than 0.05 (based on data from 16 slides) are colored in green (negative t-statistic) and red (positive t-statistic). The data are from transgenic mouse 8.*

apply Chen et al. (1997), Newton et al. (2001), and Churchill (2000)* single-slide methods to one slide from the SR–BI experiment. The different methods are used to identify genes with differential expression in mRNA samples from individual treatment mice compared to pooled mRNA samples from control mice. Using an MA–plot, Figure 2.5 displays the contours for the posterior odds of change in the Newton et al. (2001) method, the upper and lower limits of the Chen et al. (1997) 95% and 99% "confidence intervals" for M, and the contours for Sapir and Churchill's 90%, 95%, and 99% posterior probabilities of differential expression. The regions between the contours for the Newton et al. (2001) method are wider for low and high intensities A, this being a property of the Gamma distribution which is used in the hierarchical model. The genes identified as having differential expression between the SR–BI transgenic and the wild-type on the basis of all 16 slides (Callow et al., 2000) are highlighted in green (down-regulated) and red (up-regulated) in the figure. None of the methods satisfactorily identify all 13 genes found using all of the data on this

* Note that we are not performing the orthogonal regression for the log transformed intensities (Part I of the poster). The orthogonal residuals of Sapir and Churchill are essentially normalized log expression ratios. We have simply implemented Part II of the poster and are applying the mixture model to our already normalized log-ratios.

slide, and the nature of their failure strongly suggests that these methods should not be relied upon in general. In our view, the statistical assumptions the different methods make are just too strong, and inconsistent with the data, being unlikely to capture the systematic and random variability inherent these data. Furthermore, it is hard to see how a within-slide error model can capture between-slide variation, and error probabilities relating to detection of differentially expressed genes should relate to repeated hybridizations.

No single-slide or single-chip comparison exists for radiolabeled target hybridized to cDNA spots on nylon membranes or for high density short oligo arrays (i.e., for the single color systems). In order to compare mRNA from two cell populations in these cases, we need at least two nylon membranes or two chips, including at least one with target mRNA from each of the populations of interest. Assuming that we have exactly one membrane or chip from each of the cell samples, the problem becomes formally quite similar to the single-slide, two-color problem just discussed, though the details differ in important ways between nylon membranes and for high-density short oligo chips. With exactly two nylon membranes the situation really is quite similar to a single two-color slide, in that we have no more to go on than the two log intensities for each spot. Thus, determining differentially expressed genes can be no more than drawing lines in the associated plane, and the previous discussion applies, though not all the methods mentioned have been advocated in the nylon filter context. With high-density short oligo arrays, the situation is better; approximately 11 to 20 probes for each gene or EST. So there is information that permits us to estimate a standard error for each estimated log-ratio, or to carry out a significance test. Details of the Affymetrix methods for comparing two chips can be found at the following Web site: `http://www.affymetrix.com/products/software/specific/mas.affx`. This approach works reasonably well in practice, though it is not clear that the p-values can be given their usual interpretation.

2.3.3 Replicate slides: design

Before considering methods for identifying differentially expressed genes involving replicate slides, let us briefly discuss the design question for the simple treatment-control comparison with two-color arrays.

Consider the two designs described in Figure 2.2. The goal of both designs is to compare two target samples T and C, and identify differentially expressed genes between them. Suppose that we plan on doing two hybridizations, and that quantity of RNA is not a limiting factor. For a typical gene on a slide, we denote the intensity value for the two target samples by T and C. The log base 2 transformation of these values will be written $\log_2 T$ and $\log_2 C$, respectively, and when reference samples R and R' are used, we will write $\log_2 R$ and $\log_2 R'$. In addition, we denote the means of the log-intensities across slides for a typical gene by $\alpha = \mathbb{E} \log_2 T$ and $\beta = \mathbb{E} \log_2 C$, respectively. Then, for the gene under study, $\phi = \alpha - \beta$ is the parameter representing the differential expression between samples T and C, which we want to estimate.

The variances and covariances of the log-intensities for a typical gene across slides will be assumed to be the same for all samples, that is, we suppose that differential gene expression is exhibited only through mean expression levels, and we always view this on the log scale. In addition, we assume for the moment that the replicate measurements on different slides are independent. For any particular gene, let us assume that σ^2 is associated with the variance for one such measurement. (This may vary from gene to gene.) It follows that the *direct* estimate of the differential expression and its corresponding variance are:

$$\hat{\phi_D} = \frac{1}{2}(\log_2(T/C) + \log_2(T'/C')) \text{ and } var(\hat{\phi_D}) = \sigma^2/2,$$

respectively. Alternatively, if we make use of a common reference R say, then from our two hybridizations, the *indirect* estimate of log-ratio and its variance are:

$$\hat{\phi_I} = \log_2(T/R) - \log_2(C/R') \text{ and } var(\hat{\phi_I}) = 2\sigma^2.$$

The resulting relative efficiency of the indirect vs. the direct design for estimating $\alpha - \beta$ is thus 4. This is the key difference between direct and indirect comparisons, and the reason we recommend under many circumstances that direct comparisons are to be preferred. The factor 4 depends critically on our independence assumption, but we will see shortly that under very general assumptions, the direct comparison is never less precise than the indirect one.

2.3.4 Replicate slides: direct comparisons

A number of approaches can be used here, and we briefly discuss each of them.

Classical

Suppose that we have n replicate hybridizations between mRNA samples A and B. For each gene, we can compute the average $\bar{M}$ and the associated variance s^2 of the n log-ratios $M = \log_2 A/B$. In line with the early work summarized above, it would be natural to identify differentially expressed genes by taking those whose values $|\bar{M}|$ exceed some threshold, perhaps one determined by the spread of $\bar{M}$ values observed in related self-self hybridizations. It would be equally natural to statisticians to calculate the t-statistic $t = \sqrt{n}\bar{M}/s$, and make decisions on differential expression on the basis of $|t|$. Both strategies are reasonable, the first implicitly assigning equal variability to every gene, the second explicitly permitting gene-specific variances across slides; however, neither strategy is entirely satisfactory on its own. Large values of $\bar{M}$ can be driven by outliers, as the value of n is typically quite small (in our experience $\approx$ 2 to 8), and the technology is quite noisy. On the other hand, with tens of thousands of $|t|$ statistics, it is always the case that some are quite large in comparison with the others because their denominators s are very small, even though their numerators $\sqrt{n}|\bar{M}|$ may also be quite small, perhaps almost zero.

Empirical Bayes

Several more or less equivalent solutions are available to the problem of very small variances giving rise to large t-statistics, ones which lead to a compromise between solely using t and solely using $\bar{M}$. One solution is to discount genes with a small $\bar{M}$ whose standard errors are in the bottom 1%, for example. This leaves open the choice of cutoffs on $\bar{M}$ and the standard error. More sophisticated solutions in effect standardizes $\bar{M}$ by something midway between a common and a gene-specific standard error. For example, Efron et al. (2000) slightly tune the t-statistic by adding a suitable constant to each standard deviation, using

$$t^* = \frac{\sqrt{n}\bar{M}}{a + s}.$$

One choice for a is the 90th percentile of standard deviations, while another minimizes the coefficient of variation (see Efron et al., 2000; Tusher et al., 2001). This solution recalls the empirical Bayes (EB) approach to inference, which is natural in the microarray context where thousands of genes exist. A more explicitly EB approach, which is almost equivalent to the preceding one, apart from the choice of a, is presented in Lönnstedt and Speed (2001); we illustrate it next. In addition to these two just cited, there are other EB formulations of the problem of identifying differentially expressed genes (see e.g., Efron et al., 2001; Long et al., 2001; Baldi and Long, 2001; Efron and Tibshirani, 2002).

In Lönnstedt and Speed (2001), data from all the genes in a replicate set of experiments are combined into estimates of parameters of a prior distribution. These parameter estimates are then combined at the gene level with means and standard deviations to form a statistic B, which is a Bayes log posterior odds for differential expression. B can then be used to determine if differential expression has occurred. It avoids the problems of the average M and the t-statistic just mentioned. In the same article, a comparison is conducted between the B-statistic, the previous statistics t^*, and truncation of small standard errors. The differences are not great.

Note that the preceding analysis, and others like it, treats M values from different genes as conditionally independent, given the shared parameters, which is very far from the case, although it may not matter much at this point. Any attempt to provide a semiformal analysis of replicated log-ratios may require this assumption, or something very similar to it. By contrast, the permutation-based analyses described next do not require this assumption, but they do not apply here.

Robustness

The EB solution to the problems outlined in the classical approach focus on smoothing the empirical variances of the genes, thereby avoiding the situation where tiny variances can create large t-statistics. A different approach is to replace $\bar{M}$ in the numerator, in effect avoiding the problems which result from outliers coupled with small sample sizes. The obvious solution is to use a robust method of estimation of the parameter ϕ wherever possible (Huber, 1981; Hampel et al., 1986; Marazzi, 1993). Alternatively, use could be made of a nonparametric testing procedure, for example one based on ranks. The problem here is that the sample sizes are small, not infrequently as low as two or three.

Let us note in passing that the use of robust estimates of location with sample sizes as small as two or three is not without its problems. In such cases, estimates of standard errors can hardly be relied upon, and instability in the associated "t-statistics" is not uncommon. Users of standard robust procedures such as `rlm` in R and SPlus should take care and not simply rely upon default parameter settings in the algorithms. It is to be hoped that the greatly increased use of robust methods stimulated by microarray data will lead to further research on this topic.

Mixture models

Many approaches to identifying differentially expressed genes in this context use mixture models, including the EB ones just discussed. Lee et al. (2000) use a two-component normal mixture model for log-ratios in two-color arrays, one component for differentially expressed genes and another for the remainder of the genes. They estimate the parameters of their model by maximum likelihood and compute posterior probabilities using the estimated parameters. In a sense, this is another EB model, but not one with gene-specific variances, and so quite different in character from those discussed previously. Efron et al. (2001) also have a two-component mixture model, but it is for Affymetrix GeneChip data, and it is more general that the previous one. Their mixture model is for the statistic t^* instead of for log-ratios, and they make no parametric assumptions about their mixture components. More recently, Pan (2002) discussed a multicomponent normal mixture model, extending the analysis of Lee et al. (2000) toward that of Efron et al. (2001).

Fixed and random effects linear models

One approach to the analysis of two-color microarray data and the determination of differentially expressed genes makes use of linear models and the analysis of variance (Kerr et al., 2000; Wolfinger et al., 2001; Jin et al., 2001). These authors model unnormalized log intensities with linear models which include terms for slide, dye, gene and treatment, a subset of the interactions between these effects, and a random error term. Important differences exist between the approaches of Kerr et al. (2001) and the other two papers in the way in which normalization is incorporated, in whether terms are fixed or random, and in assumptions about the error variances. We illustrate the approach of Kerr et al. (2001) next, and we begin by presenting their model for our Case study I. We label array effects (A) by i, dye effects (D) by j, treatment effects (V) by k and gene effects (G) by g. Their model for the log of the intensity y_{ijkg} is:

$$\log(y_{ijkg}) = \mu + A_i + D_j + V_k + G_g + (AG)_{ig} + (VG)_{kg} + \epsilon_{ijkg},$$

where $i = 1, \ldots, 4$, $j = 1, 2$, $k = 1, 2$, and $g = 1, \ldots, 8848$. In case study I, which consists of a two sample comparision, the treatment effect V_k represents the mutant

(V_2) or wild-type (V_1) samples. In this model, all terms are fixed apart from the random error terms ϵ. The term of interest here is $(VG)_{kg}$ and the value $(VG)_{2g} - (VG)_{1g}$ estimates the level of differential expression between the mutant and the wild type samples for gene g. This model can easily be extended to cover multiple samples and factorial designs, although we do it differently in Section 2.4.3. Further discussion of fixed and random effects linear model is given there.

Error models

Several groups have used more fully developed error models for measurements on microarrays, and sought to identify differentially expressed genes by making use of their error model. These include Roberts et al. (2000), Ideker et al. (2000), Rocke and Durbin (2001), Theilhaber et al. (2001), and Baggerly et al. (2001). Ideker et al. (2000) has made their software publicly available, so we illustrate this approach in Case Study I. Their error model is for pairs T and C of unnormalized unlogged intensities for the same spot, namely,

$$\begin{aligned} T &= \mu_T + \mu_T \epsilon + \delta \\ C &= \mu_C + \mu_C \epsilon' + \delta', \end{aligned}$$

where $(\epsilon,\ \epsilon')$ and $(\delta,\ \delta')$ are independently bivariate normally distributed across spots with means $(0, 0)$ and separate, general covariance matrices that are common to all spots on the array. Thus, six parameters exist for the error model, in addition to parameters determining the expected values. This model is close, but not quite identical to a similar model for $(\log T, \log C)$, one that permits the two components to be correlated, with a correlation which is common to all the spots. Intensities from distinct spots on the same slide, or spots on different slides are taken to be independent. Roberts et al. (2000) and Rocke and Durbin (2001) use a similar model, but do not allow the different channel intensities to be correlated.

In order to identify differentially expressed genes, Ideker et al. (2000) fit the model to two or more slides by maximum likelihood. Differential expression is then determined by carrying out a likelihood ratio test of the null hypothesis $\mu_T = \mu_C$ for each gene separately, resulting in a likelihood ratio test statistic λ for each gene.

Other approaches

Working with Affymetrix GeneChip data, Thomas et al. (2001) use a variant on the simple two-sample t-statistic which starts from a nonlinear model including sample-specific additive and multiplicative terms. In the same GeneChip context, Theilhaber et al. (2001) present a fully Bayesian analysis, building on a detailed error model for such data. We refer readers to these articles for further details.

2.3.5 Replicate slides: indirect comparisons

When the comparisons of log-ratios in a two-color experiment are all indirect, the procedures just described need to be modified slightly. Typically, we would have some number, n_T, of slides on which the gene expression of sample T is compared to that of a reference sample R, leading to n_T log-ratios $M = \log_2 T/R$, and a similar set of n_C log-ratios $\log_2 C/R'$ from sample C. The analogues of the average and t-statistics here are the differences between the means $\bar{M_T} - \bar{M_C}$ and the two-sample t-statistic is

$$t = \frac{\bar{M_T} - \bar{M_C}}{s_p\sqrt{1/n_T + 1/n_C}},$$

where s_p is the pooled standard deviation.

The problems with these two statistics are completely analogous to those described in the previous section with $\bar{M}$ and t, and the solutions are similar: modify s_p or use robust variants of $\bar{M_T} - \bar{M_C}$ and t.

We note in closing this brief review that some non-statisticians addressing these two-sample problems in the microarray literature have devised novel approaches. Galitski et al. (1999) and Golub et al. (1999) sought to identify single differentially expressed genes by computing for each gene the correlation of its expression profile with a reference expression profile, such as a vector of indicators for class membership. In the case of two classes, this correlation coefficient is a type of t-statistic. Genes were then ranked according to their correlation coefficients, with a cutoff derived from a permutation distribution.

2.3.6 Illustrations using our case studies

We now describe the results of applying some of the methods just discussed to case studies I and II. It should be understood that we are not attempting to present thorough analyses of these datasets, in part because of lack of space, and in part because more effort always goes into the determination of differential expression than the application of a single statistical analysis. Nevertheless, we hope that what follows gives an indication of the potential of the methods we illustrate. We have certainly found them useful in similar contexts, particularly the graphical displays.

Plots: averages, SDs, t-statistics and overall expression levels

Important features of the genes which might be differentially expressed can be found by examining plots of t-statistics, their numerators $\bar{M}$ and denominators s, and the corresponding overall expression levels. The overall expression level for a particular gene is conveniently measured by the quantity $\bar{A}$, the average of $A = \log_2 \sqrt{RG}$ over all the slides in the experiment.

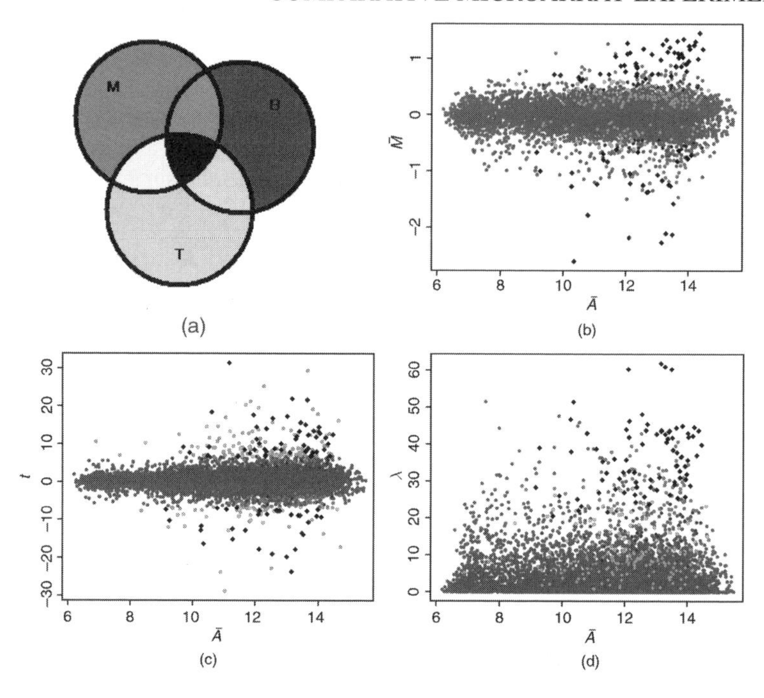

Figure 2.6 *(See color insert following page 114.) Graphical illustration of case study I: swirl experiment. (a) The color code that defines three groups of 250 genes consisting of largest values of* $|\bar{M}|$, $|t|$ *and* B *values; (b)* $\bar{M}\bar{A}$*–plot; (c)* t *vs.* $\bar{A}$*; (d)* λ *vs.* $\bar{A}$.

Let us begin with the *swirl* experiment. In Figure 2.6b, Figure 2.6c and Figure 2.6d, we have plotted the average log ratio $\bar{M}$, the t-statistic and the log-likelihood ratio statistic λ from Ideker et al. (2000) against $\bar{A}$. We began by defining three groups of 250 spots each, being those having the largest values of $|\bar{M}|$, $|t|$ and B, respectively, and then plotted the points according to the color code depicted in Figure 2.6a. Thus, points corresponding to spots in all three groups are colored heavy black, while the very light black spots, the overwhelming majority, are in none of the groups. Clearly, the heavy black spots are the main candidates for differential expression. Spots belonging to the large $|t|$ group only are green, while those in the large $|\bar{M}|$ and large B and not large $|t|$ groups are pink. The absence of points colored yellow is noteworthy: any spots with large $|t|$ and large $|\bar{M}|$ already have large B. This is not uncommon, and shows that B is, in general, a useful compromise between t and $\bar{M}$.

It is clear from these plots, and is not infrequently the case, that some points are well separated from the cloud. The genes corresponding to these points are likely to be differentially expressed, and we recommend this informal approach identifying such genes. In many cases, this evidence is as solid as any we are able to obtain for differential expression. Note the broad agreement on the black points in all three of the panels. Figure 2.7a shows the nature of B rather clearly, while Figure 2.7b shows the role of the standard deviation (SD) in determining whether a given $\bar{M}$ ends up

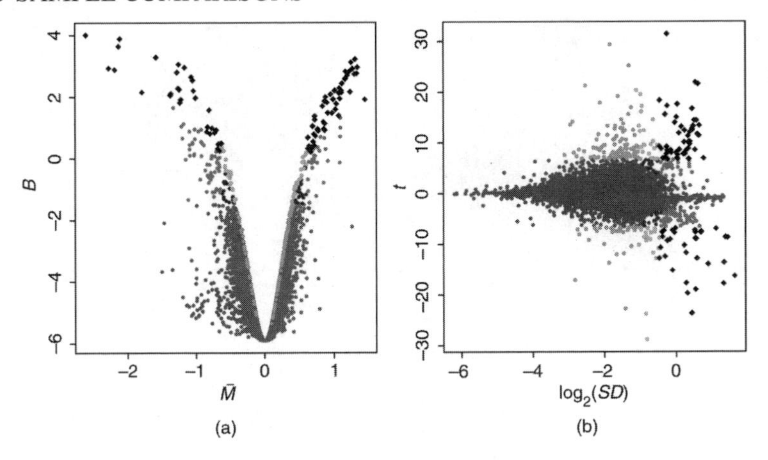

Figure 2.7 *(See color insert following page 114.) Case study I: swirl experiment. Plots of (a) Log odds B vs.* $\bar{M}$ *and (b) t vs.* $\log_2(SD)$. *Spots (genes) corresponding to large* $|\bar{M}|$, $|t|$ *and B values highlighted according to color code shown in Figure 2.6a.*

having a large $|t|$ as well, with it being clear how large B related to the other two. The green points are those with smaller SDs, in comparison with the solid black, pink, and light blue spots. Of course, much of this is dependent on the cutoffs defining our groups, which were determined after looking at the plots, but the message of these plots is general.

Turning to the SR–BI experiment, Figure 2.8a shows a plot of $\bar{M_T} - \bar{M_C}$ against $\bar{A}$, while Figure 2.8b shows t against $\bar{A}$. Here, we have given no analogue of B, though one could be developed. In this case it is the yellow spots that should attract our attention, and it should be clear that our choice here of 250 in the two groups is too large. Looking at Figure 2.9a and 2.9b, we see that in general a large $|\bar{M_T} - \bar{M_C}|$ is more likely to go with a large $|t|$ if the SD is not too large and not too small, which

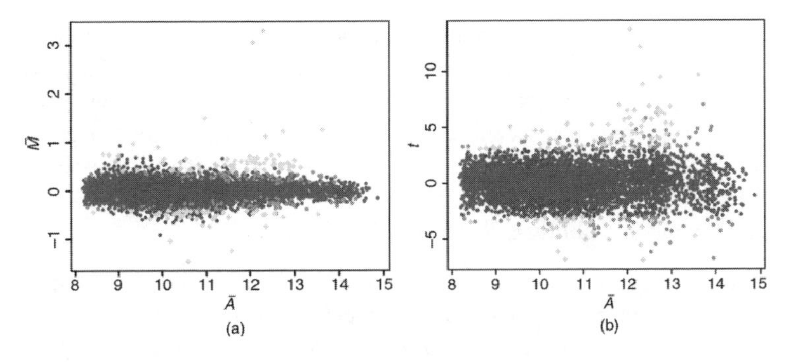

Figure 2.8 *(See color insert following page 114.) Case study II: SR-BI experiment: (a)* $\bar{M}\bar{A}$*–plot; (b) t vs.* $\bar{A}$. *Spots (genes) corresponding to large* $|\bar{M}|$ *and* $|t|$ *values highlighted according to color code shown in Figure 2.6a.*

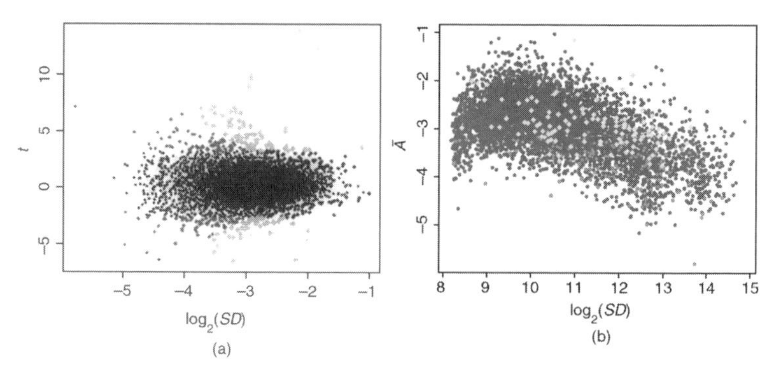

Figure 2.9 *(See color insert following page 114.) Case study II: SR–BI experiment. (a)* t *vs.* $\log_2(SD)$*; (b)* $\log_2(SD)$ *vs.* $\bar{A}$

is roughly equivalent to having an $\bar{A}$ value that is not too small and not among the largest ones. We see clearly from Figure 2.9b that larger $\bar{A}$ values go with smaller values of the SD, and we remark that many plots of this kind would show a much greater increase in SD for lower values of $\bar{A}$. That is often quite evident in the $\bar{M}\bar{A}$-plot (Figure 2.8a) as a ballooning in the low intensity ranges, a phenomenon we have prevented by using a smaller and less variable background adjustment (see Yang et al., 2002a).

How would we use these statistics to identify differentially expressed genes? It is tempting to answer in the following way: Simply rank the genes on the basis of $|\bar{M}|$, t or B, and determine a cutoff in some sensible way. This is in effect what we have done in producing the plots, without being careful about cutoffs; however, the plots also tell us that a ranking based on just one of these statistics is not necessarily the best we can do. The reason is this: A spot's overall intensity value $\bar{A}$ can be a useful indicator of the importance that can be attached to its $\bar{M}$, t, or B value. Typically, considerably fewer genes exist with large $\bar{A}$ values, and a spot can stand out from the cloud in the larger $\bar{A}$ range with a smaller $\bar{M}$, t, or B value than would be necessary to stand out in the low $\bar{A}$ region. At times, the difference can be striking, although this is not the case with our illustrative examples. Most methods of identifying differentially expressed genes in the present context do not make explicit use of the overall intensity or any related values, although approaches using fully specified error models in effect do so. One problem with full error models is that they invariably assume that observations on different slides are independent, which can be very far from the case, and this vitiates their exact probability calculations.

In summary, we regard the determination of differentially expressed genes on the basis of a small number of replicates as a problem for which more research is needed. It is clear that the values of $\bar{M}$, t, B (or their analogues) are highly relevant, but the values of $\bar{A}$ and the SD should not be ignored. For the present moment, we feel that determining cutoffs is best done informally, following visual inspection of plots like the ones we have shown. Naturally, scientific considerations such as the expected number of differentially expressed genes and the number of follow-up experiments that might be feasible, are also relevant in determining cutoffs, and we feel that $\bar{A}$ and other similar statistics should also play a role.

Ranking genes by sets of statistics

Whether we use of $|\bar{M}|$, $|t|$, B, λ or some similar statistic to provide a ranking of genes corresponding to the strength of evidence of differential expression, it is intuitively clear that our preferred genes should rank highly on all these criteria. Hero and Fleury (2002) describe a valuable method of selecting genes which are highly ranked in a suitable multidimensional sense. They describe what they term *Pareto fronts* in *multicriterion scattergrams*, which are points that are maximal in the componentwise ordering (Pareto optimal) in the P-dimensional scatterplots of a desired set of P criteria. As well as giving some modified versions of Pareto optimal genes, they demonstrate that their ideas are applicable to a wide range of gene filtering tasks, not just that of detecting differentially expressed genes. We refer to the paper Hero and Fleury (2002) and also Fleury et al. (2002b) for a fuller discussion of the method.

2.3.7 Assessing significance

After ranking the genes based on a statistic, a natural next step is to choose suitable cutoff values defining the genes that might be considered as significant or differentially expressed. In this section, we consider the extent to which this can be done informally. We focus on two types of plots.

Quantile–quantile plots

A simple graphical approach is to examine the *quantile–quantile plots* (Q–Q plots) for certain statistics. Q–Q plots are a useful way to display the $\bar{M}$ or t-statistics for the thousands of genes being studied in a typical microarray experiment. The standard normal distribution is the natural reference, and the more replicates that enter into an average or t-like statistic, the more we can expect the majority of the statistics to look like a sample from a normal distribution. With just four replicates, which is equivalently eight log intensities entering into our averages, we cannot expect and do not get a very straight line against the standard normal; however, the plot can have value in indicating the extent to which the extreme t-statistics diverge from the majority. Q–Q plots informally correct for the large number of comparisons, and the points which deviate markedly from an otherwise linear relationship are likely to correspond to those genes whose expression levels differ between the two groups. At times, we can tell where the outliers end and the bulk of the statistics begin (see e.g., Dudoit et al., 2001b), but unfortunately this is not the case with either of the present examples.

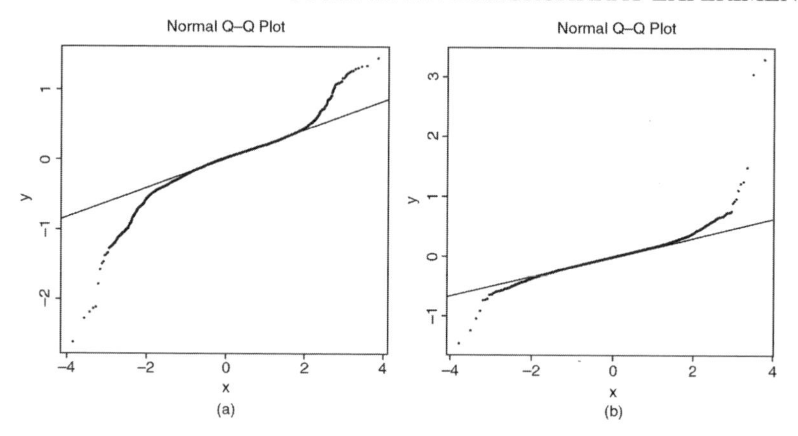

Figure 2.10 *Quantile-quantile (Q–Q) plot for (a) one-sample t-statistics from the swirl experiment; (b) two-sample t-statistics from the SR–BI experiment.*

Figure 2.10a is typical of many Q–Q plots which are not very helpful. The tails of the distrubutions of these t-statistics are far from those of a normal (or t) distribtion. The best we can hope for in such cases is that some points are obviously outliers with respect to the nonnormal distribution, and this is true to some extent here, especially for large negative ts.

The picture in Figure 2.10b is clearer. This is for the SR–BI experiment, and is perhaps what we would expect, as there are 16 observations in each of these t-statistics. We see about a dozen genes with "unusual" t-statistics, and these are obviously good candidates for genes exhibiting differential expression (both up and down-regulation). Some of these genes were verified to have the observed behavior in follow-up experiments.

p-value vs. average M (volcano) plot

Another plot which allows outliers to reveal themselves among thousands of statistics is the so-called volcano plot (Wolfinger et al., 2001), Figure 2.11. We have already seen one plot like this in Figure 2.7a, where the log-odds corresponding to a given value of the statistic B was plotted against $\bar{M}$. More commonly, people plot the logs of raw (i.e., model-based, unadjusted) p-values against the estimated fold change on a log scale, $\bar{M}$ (Wolfinger et al., 2001). Whether the p-values are calculated assuming a t or a normal distribution is not so important here. The color code indicates how these plots capture some aspects of the plots we presented earlier in what is perhaps a more convenient form; see especially the solid black points in Figure 2.11a and the yellow points in Figure 2.11b.

So far, the approaches we have offered for identifying differential expression are all informal. In many cases, this will be adequate. The number of genes that are selected as possibly differentially expressed will in general depend on many things: the aims

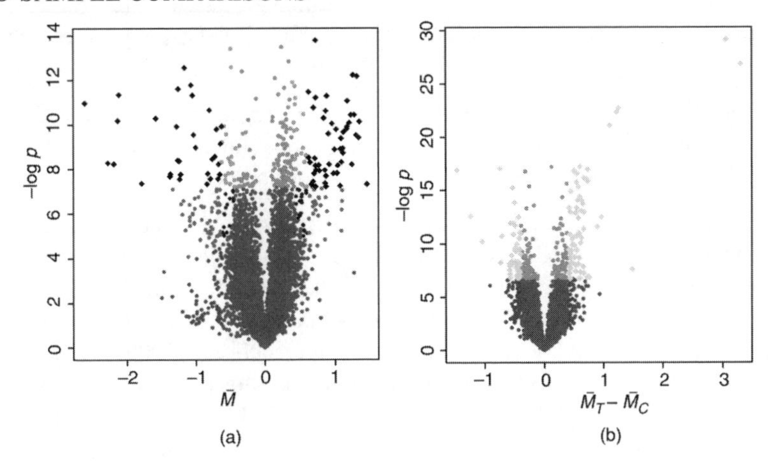

Figure 2.11 *(See color insert following page 114.) (a) Case study I: swirl experiment:* $-\log p$*-value vs. average log-fold change* ($\bar{M}$)*. (b) Case study II: SR-BI experiment:* $-\log p$*-value vs. difference of log fold change* $\bar{M}_T - \bar{M}_C$*. Color code as in Figure 2.6a.*

of the experiment, the number of genes expected to be differentially expressed, the number of replicates, and the nature and extent of follow-up (validation). We cannot expect a formal procedure to take all these factors into account.

In order to examine more formally whether the extreme t-statistics do indeed reflect real differences between the control and transgenic mice we turn to adjusted p-values. These cannot always be calculated in a reliable way, but when they can, we feel they are worthwhile.

2.3.8 Multiple comparisons

A more formal approach to testing the null hypothesis of constant expression can be obtained by calculating p-values (or posterior probabilities) under some model. With a typical microarray dataset comprising thousands of genes, however, there are at least two major impediments to doing this correctly. One immediate concern for the validity of any single gene unadjusted p-values or posterior probabilities is the underlying statistical model. As we suggested, in our discussion of single-slide methods for identifying differentially expressed genes, parametric statistical models generally have difficulty adequately capturing all the details of microarray data; this is worst where it matters most for testing, namely, in the tails of the distributions of statistics. For example, collections of single-slide log-ratios M are frequently approximately normally distributed, but this breaks down in the tails. We might hope for averages to behave more like normal random variables by virtue of the central limit theorem, but this will only be true when all systematic effects have been removed and the individual terms really do exhibit a high degree of independence. In certain cases, permutation-based methods can be used, but this is not as frequent as we might like. As we will see, use of permutation methods can raise yet other problems.

The other major concern is the multiplicity issue: Carrying out thousands of significance tests (or computing thousands of posterior probabilities) brings with it the need to recognize and deal with the issues that arise when so many comparisons are made.

Two different approaches to the multiple comparisons problem have emerged in the microrarray literature. One makes use of the traditional literature on the problem, seeking to control family-wise type I error rates (see Dudoit et al., 2002b), who build on the work of Westfall and Young (1993), while the other develops and extends the notion of false discovery rate of Benjamini and Hochberg (1995) (see e.g., Efron et al., 2000; Tusher et al., 2001). We summarize and illustrate both approaches and refer to Dudoit et al. (2002a) and Storey et al. (2002) for fuller details and further references. We also refer to Manduchi et al. (2000) for a contribution to this problem which falls outside the approaches just mentioned.

Consider the problem of simultaneously testing m null hypotheses $H_j, j = 1, \ldots, m$, and denote by R the number of rejected hypotheses. In the frequentist setting, the situation can be summarized by the following table (Benjamini and Hochberg, 1995). The m hypotheses are assumed to be known in advance, while the numbers m_0 and m_1 of true and false null hypotheses are unknown parameters, R is an observable random variable, and S, T, U, and V are unobservable random variables. In the microarray context, a null hypotheses H_j exists for each gene j, and rejection of H_j corresponds to declaring that gene j is differentially expressed. In general, we want to minimize the number V of *false positives*, or *type I errors*, and the number T of *false negatives*, or *type II errors*. The standard approach is to prespecify an acceptable type I error rate α and seek tests that minimize the type II error rate (i.e., maximize *power*), within the class of tests with type I error rate α.

In terms of these random variables, we can define the main rates used in the present context. The *per-comparison error rate* (*PCER*) is defined as the expected value of the number of type I errors over the number of hypotheses (i.e., $PCER = \mathbb{E}(V)/m$. The *family-wise error rate* (*FWER*) is the probability of at least one type I error, $FWER = pr(V \geq 1)$. The *false discovery rate* (*FDR*) is the expected proportion of type I errors

	No. not rejected	No. rejected	
No. true null hyptheses	U	V	m_0
No. nontrue null hyptheses	T	S	m_1
	$m - R$	R	m

among rejected hypotheses, $FDR = \mathbb{E}(V/R; R > 0) = \mathbb{E}(V/R|R > 0)pr(R > 0)$, in Benjamini and Hochberg (1995), while the *positive false discovery rate* (*pFDR*) in Storey (2002) is the rate that discoveries are false, $pFDR = \mathbb{E}(V/R|R > 0)$.

It is important to note that the preceding expectations and probabilities are *conditional* on which hypotheses are true or false (i.e., on which genes are differentially expressed). We distinguish between strong and weak control of the type I error rate. *Strong control* refers to control of the type I error rate under any combination of true and false hypotheses (i.e., for any combination of differentially and constantly expressed genes). In contrast, *weak control* refers to control of the type I error rate only when none of the genes are differentially expressed (i.e., under the *complete null hypothesis* H_0^C that all the null hypotheses are true). In general, weak control without any other safeguards is unsatisfactory. In the microarray setting, where it is very unlikely that none of the genes are differentially expressed, it seems particularly important to have strong control of the type I error rate.

As we will see, a wide variety of multiple testing procedures are used. How should we choose which to use? No simple answers are available here, but a procedure might be judged according to a number of criteria. One criterion is interpretation: Does the procedure answer a question that is relevant to the investigation? Another is type of control: strong or weak? We have already suggested that strong control is highly desirable in the microarray context. An important criterion is validity: Are the assumptions under which the procedure applies clearly true, or perhaps plausibly true, or are they unclear, or most probably not true? A fourth is computability: Are the procedure's calculations straightforward to carry out accurately? Or, is there perhaps numerical or simulation error, or discreteness, which might cast doubt on the exactness of the result?

Adjusted p-values

To account for multiple hypothesis testing, one may calculate adjusted p-values (Westfall and Young, 1993). Given a test procedure, the *adjusted p-value* corresponding to the test of a single hypothesis H_j can be defined as the level of the entire test procedure at which H_j would just be rejected, given the values of all test statistics involved. We can distinguish three ways of adjusting p-values: the single-step, step-down, and step-up procedures. In *single-step* procedures, equivalent multiplicity adjustments are performed for all hypotheses, regardless of the ordering of the test statistics or unadjusted p-values. Improvement in power, while preserving type I error rate control, may be achieved by *stepwise procedures*, in which rejection of a particular hypothesis is based not only on the total number of hypotheses, but also on the outcome of the tests of other hypotheses. *Step-down* procedures order the unadjusted p-values (or test statistics) starting with the most significant, while *step-up* procedures start with the least significant.

For strong control of the FWER at level α, the Bonferroni procedure rejects any hypothesis H_j with p-value less than or equal to α/m. The corresponding *Bonferroni single-step adjusted p-values* are thus given by $\tilde{p}_j = \min(mp_j, 1)$. While single-step adjusted p-values are simple to calculate, they tend to be very conservative. Let the *ordered unadjusted p-values* be denoted by $p_{r_1} \leq p_{r_2} \leq ... \leq p_{r_m}$. Then, the *Holm step-down adjusted p-values* are given by

$$\tilde{p}_{r_j} = \max_{k=1,\ldots,j} \left\{ \min\left((m-k+1)\, p_{r_k}, 1\right) \right\}.$$

Holm's procedure is less conservative than the standard Bonferroni procedure which multiplies the p-values by m at each step. However, neither Holm's method nor Bonferroni (nor other single-step methods) take into account the dependence between the test statistics, which can be quite strong for co-regulated genes. Westfall and Young (1993) propose adjusted p-values which take into account quite general dependence, and which are less conservative. Their *step-down minP adjusted p-values* are defined by

$$\tilde{p}_{r_j} = \max_{k=1,\ldots,j} \left\{ pr\left(\min_{l \in \{r_k,\ldots,r_m\}} P_l \leq p_{r_k} \mid H_0^C \right) \right\},$$

while their *step-down maxT adjusted p-values* are defined by

$$\tilde{p}_{s_j} = \max_{k=1,\ldots,j} \left\{ pr\left(\max_{l \in \{s_k,\ldots,s_m\}} |T_l| \geq |t_{s_k}| \mid H_0^C \right) \right\},$$

where $|t_{s_1}| \geq |t_{s_2}| \geq ... \geq |t_{s_m}|$ denote the ordered test statistics.

These adjusted p-values lead to procedures which guarantee weak control of the FWER in all cases, and strong control under the additional assumption of subset pivotality, which applies in the example described next. See Dudoit et al. (2002a) for fuller details, including details of permutation-based calculation of the p-values and their step-down adjustments.

Turning now to a different kind of adjusted p-value, the following formula of Benjamini and Hochberg (1995) gives a step-up adjustment which leads to strong control of the FDR under the additional assumption of independence of the test statistics. Even though this is not a realistic assumption with microarray data, we nevertheless offer the formula

$$\tilde{p}_{r_i} = \min_{k=i,\ldots,m} \left\{ \min(mp_{r_k}/k, 1) \right\}.$$

Recently, Benjamini and Yekutieli (2001a) showed that the preceding adjustment is applicable when a particular form of dependence they term positive regression dependency obtains between the test statistics, and gave a conservative modification applicable quite generally. The latter formula is equivalent, for large m, to replacing the factor m multiplying the unadjusted p-values by $m \log m$.

The final multiple comparison method we discuss briefly is the *pFDR* of Storey (2002). This novel approach does not seek to control the type I error or the false-discovery rate, nor does it provide adjusted p-values. Instead, Storey (2002) takes the view that conservatively estimating the FDR or $pFDR$ for rejection regions defined beforehand or by the actual statistics is a more appropriate task. In a separate paper, Storey (2001) introduces the notion of q-value, which loosely speaking, is the minimum $pFDR$ that can occur when rejecting a statistic equal to the observed one for a nested set of rejection regions. Space does not permit us to present full definitions or explanations of this highly interesting theory, so we refer the reader to the articles by Storey (2002, 2001). These articles are mainly concerned with the case where the test statistics are independent and identically distributed, while Storey and Ribshirani (2001) requires only identically distributed test statistics and an ergodicity condition.

Comparison of multiple testing procedures

We now apply the procedures to case study II, the SR–BI experiment, and display the results in Figure 2.12. All p-values depicted there are for the two-sample t-statistics introduced in the previous section for this example, and are calculated using all 12,870 assignments of 16 mice to two groups of 8, called "treatment" and "control." For example, the unadjusted (two-sided) p-value p_i for gene i will be 2/12,870 when none of the assignments of the 16 mice to two groups of 8, other than the true one, leads to a t-statistic larger in absolute value than t_i, the value observed with the true assignment.

Figure 2.12 presents a variety of adjusted p-values for the genes with the 100 smallest unadjusted p-values. As the solid line indicates, all of the latter are well below 0.01. The Holm (and hence the Bonferroni) adjustment (dashes) is already far too conservative, suggesting that none of the genes are differentially expressed, something we know

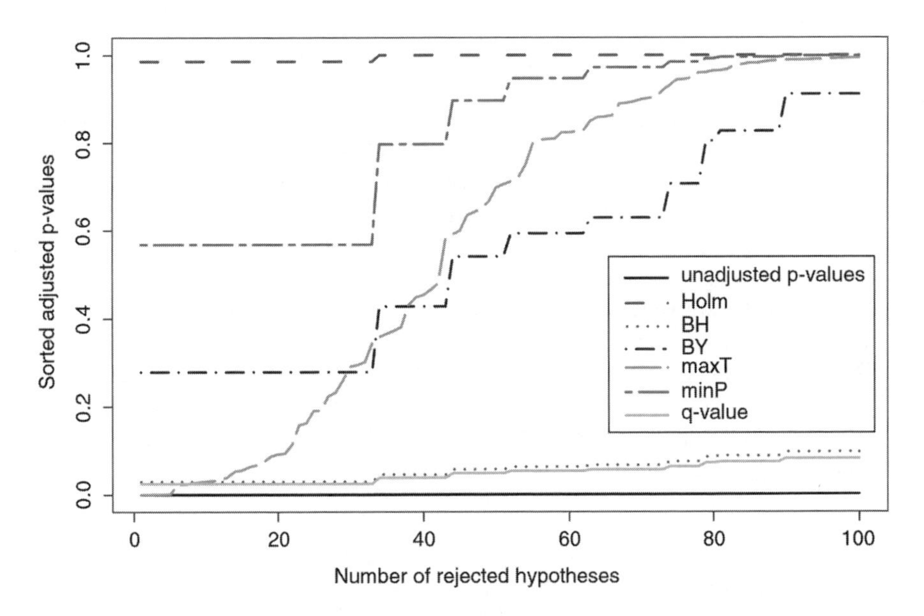

Figure 2.12 *(See color insert following page 114.) Comparisons of multiple testing procedures. The plots concern the spots with the 100 smallest unadjusted p-values, scored 1 to 100 horizontally. The different adjusted p-values and the q-values are vertical; see insert for line code.*

from follow-up studies to be false; but with over 6,000 genes and a smallest possible unadjusted p-value of roughly 1/6,000, this is not surprising. The minP adjustment (short dash, long dash) is likewise too conservative, and the explanation is the same. Even an observed t-statistic whose magnitude is not exceeded by that of any of the other (false) assignments of treatment and control status to the mice, leads to a minP adjusted p-value of 0.53, a value which is determined by the number of permutations and not by the magnitude of the test statistic. The adjustment of Benjamin and Yekutieli (2001b) (dot-dash) is similarly conservative. The one adjustment that does lead to fairly small values is the maxT adjustment (curved dash), and here we must point out that the permutation distributions of the 6000 t-statistics are not all identical, and consequently not all tests contribute equally to the maxT adjusted p-values, Westfall and Young (1993). In this example, the adjustment of Benjamini and Hochberg (1995) (which assume independence) turns out to be very close to the q-values of Storey (2001) (dots). While not giving values that are conventionally small (e.g., less than 0.05), these adjustments do seem less conservative than the others described previously.

In closing this brief discussion of multiple testing, we draw some tentative conclusions. The well-established and generally applicable p-value adjustments such as Bonferroni's, Holm's, and Westfall and Young's minP do not seem to be helpful in the microarray context. Although 12,870 permutations might seem rather few in some contexts, 8 replicates of a treatment and 8 of a control are large numbers for a microarray experiment: 2, 3, and occasionally 5 are more common numbers of replicates. Also, it is not easy to argue that there should be more, simply to permit the use of this multiple testing methodology. In many, perhaps most cases, 3–5 replicates will be enough to reveal many differentially expressed genes. Thus, we are unlikely to see many situations in which permutation based (and hence distribution-free) adjusted p-values can be used. Westfall and Young's maxT appears somewhat more useful. Procedures based on controlling or estimating the FDR (or pFDR) appear to be the most promising alternatives. When thousands of tests are being carried out, concern about making one or more false positive decisions does seem misplaced. Instead, permitting a small percentage of false positives among the rejected hypotheses seems more reasonable. However, the theory for FDR and pFDR needs to be developed more before it is generally applicable to microarray data. We need procedures controlling FDR under fairly general dependence assumptions between the test statistics that are less conservative than the one presented earlier. Also, the theory for pFDR needs to be applicable beyond independent or identically distributed test statistics in a framework that definitely covers microarray data.

2.3.9 Correlation and technical replicates

As explained in Section 2.3.3, an important decision with two-color microarray experiments when choice is available is whether to use direct or indirect comparisons, that is, whether to measure expression differences within slides or between slides.

Most statisticians would immediately answer that direct comparisons are better, and they are, but the situation is not quite as simple as it might look upon first examination.

In Section 2.2.4, we explained how biologists conducting simple gene expression comparisons, such as between treated and control cells, will typically carry out replicate experiments on different slides. These will frequently involve what we have called *technical replicates*, where the target mRNA in each hybridization is from the same RNA extraction, but is labeled independently for each hybridization. We have noticed that estimates of differential gene expression based on technical replicates tend to be positively correlated, whereas the same estimates based on replicates involving different RNA extractions and labelings tend to be uncorrelated. When the more extreme form of technical replication is used, that is, when a labeled sample is split and used in replicate hybridizations, the correlation can be very strong. For scatter plots illustrating these assertions, see Figure 7-5 of Bowtell and Sambrook (2002). These observations suggest that we should reexamine the independence assumption underlying the experimental design calculations presented in Section 2.2.3. When variances are calculated for linear combinations log-ratios across replicate slides, it appears desirable to use the most realistic covariance model for the measurements. The discussion that follows is based on Speed and Yang (2002) to which we refer for fuller details.

A more general covariance model

Let us re-examine comparisons between two target samples T and C. Following the notation from Section 2.3.3, for a typical gene on the slide, we denote the means of the log-signals across slides by $\alpha = \mathbb{E}\log_2 T$ and $\beta = \mathbb{E}\log_2 C$, respectively. The variances and covariances of the log signals across slides will be assumed to be the same for all samples. Our dispersion parameters are a common variance τ^2, a covariance γ_1 between measurements on samples from the same hybridization, a covariance γ_2 between measurements on technical replicate samples from different hybridizations, and a covariance γ_3 between measurements on samples which are neither technical replicates nor in the same hybridization. These parameters will in general be different for different genes, but we suppress this dependence in our notation. Later, we attempt to estimate typical values for them.

Consider again the two different designs illustrated in Figure 2.2. Design I illustrates an indirect comparison, where T and C are each hybridized with a common reference sample R. Design II involves two direct comparisons, where the samples T and C are hybridized together on the same slide. We denote technical replicates of T, C, and R by T', C', and R', respectively and $log_2 T$, $log_2 C$ etc. by t, c etc. Note that both designs involve two hybridizations, and we emphasize that in both cases, our aim is to estimate the expression difference $\phi = \alpha - \beta$ on the log scale. We now calculate the variances of the obvious estimates of this quantity from each experiment. For Design I, this is

$$v_1 \;=\; var(t - r - c + r') \;=\; 4(\tau^2 - \gamma_1) - 2(\gamma_2 - \gamma_3),$$

while for Design II, the estimate is one-half of $y = t - c + t' - c'$, and we have

$$v_2 = var(y/2) = \tau^2 - \gamma_1 + \gamma_2 - \gamma_3.$$

We next show that v_2 is never greater than v_1. To see this, consider the covariance matrix for the four log intensities from Design II:

$$cov \begin{pmatrix} t \\ c \\ t' \\ c' \end{pmatrix} = \begin{bmatrix} \tau^2 & \gamma_1 & \gamma_2 & \gamma_3 \\ \gamma_1 & \tau^2 & \gamma_3 & \gamma_2 \\ \gamma_2 & \gamma_3 & \tau^2 & \gamma_1 \\ \gamma_3 & \gamma_2 & \gamma_1 & \tau^2 \end{bmatrix}.$$

It is easy to check that the eigenvalues of this matrix are $\lambda_1 = \tau^2 + \gamma_1 + \gamma_2 + \gamma_3$, $\lambda_2 = \tau^2 + \gamma_1 - \gamma_2 - \gamma_3$, $\lambda_3 = \tau^2 - \gamma_1 + \gamma_2 - \gamma_3$, and $\lambda_4 = \tau^2 - \gamma_1 - \gamma_2 + \gamma_3$, corresponding to the eigenvectors $(1, 1, 1, 1)'$, $(1, 1, -1, -1)'$, $(1, -1, 1, -1)'$ and $(1, -1, -1, 1)'$, respectively. In terms of these eigenvalues, we see that $v_1 = \lambda_3 + 3\lambda_4$ and that $v_2 = \lambda_3$. Thus, the relative efficiency of the indirect vs. the direct design for estimating $\alpha - \beta$ is

$$\frac{v_1}{v_2} = 1 + \frac{3\lambda_4}{\lambda_3}.$$

The direct design is evidently never less precise than the indirect one, and the extent of its advantage depends on the values of $\tau^2, \gamma_1, \gamma_2$, and γ_3. Notice that when $\lambda_4 = 0$ (equivalently, $\tau^2 - \gamma_1 = \gamma_2 - \gamma_3$), we see that $v_1 = v_2$. This shows that under our more general model, the reference design could, in theory, be as efficient as the direct design. This is very unlikely in practice, as these conditions are equivalent to the variance $var(t - c)$ of a log-ratio coinciding with the covariance $cov(t - c, t' - c')$ between two log-ratios derived from technical replicate samples. At the other extreme, when $\gamma_2 = \gamma_3$, that is when the covariance between measurements on technical replicates coincides with that between any two unrelated samples, we have $v_1 = 4v_2$. This is the conclusion which is obtained when log-ratios from different experiments are supposed independent.

The preceding discussion focused on a single gene. It is not an easy task to obtain estimates of these eigenvalues or of v_1 and v_2 for single genes; however, in Speed and Yang (2002), we presented an analysis based on the data from the *swirl* experiment, which sought to obtain estimates of the average eigenvalues, and corresponding estimates of averages for v_1 and v_2. We found there that the relative efficiency v_2/v_1 of the indirect to direct designs for estimating $\log_2(swirl/wt)$ was 4 for dye-swap set 1, consistent with independence, and 2.5 for dye-swap 2, suggesting a measure of dependence.

When we average log-ratios, as we do in Design I, we want the terms to be as independent as possible to minimize the relevant variance. In this case, it would be best if we could avoid using technical replicates, and use truly independent samples. On the other hand, when we take differences, as we do in Design II, we want the technical replicate terms (R and R') to be as dependent as possible. This could be achieved by using the same extraction and the same labeling (extreme technical replication) for the common reference mRNA. Further, the results we have just described suggest that, in some cases, the covariance due to technical replication needs to be considered. We will revisit this discussion with a three-level factor in the next section.

2.4 Single-factor experiments with more than two levels

A simple extension to two-sample comparisons is single-factor experiments where we wish to compare the effects of K treatments, $T_0, \ldots, T_{K-1}$, on gene expression. Examples of such experiments include comparing multiple drug treatments to a particular type of cell (Bedalov et al., 2001) or comparing different spatial regions of the retina (Diaz et al., 2002b).

2.4.1 Case study III: mouse olfactory bulb

To provide an illustration, we consider an experiment where comparisons were made between different spatial regions of mouse olfactory bulb to screen for possible region specific developmental cues (Lin et al., 2002). The target cDNA was hybridized to glass microarrays containing 18,000 cDNA probes obtained from the Japanese Institute of Physical and Chemical Research (RIKEN) consortium. The olfactory bulb is an oblong spherical structure, so in order to make a three-dimensional representation using binary comparisons, the bulb was dissected into three sections along the three orthogonal axes, leading to six samples termed anterior (front), posterior (back), medial (close to the axis of bilateral symmetry), lateral (away from the axis of symmetry), dorsal (top), and ventral (bottom). In what follows, these cell samples will be denoted by A, P, M, L, D, and V, respectively. Initially, comparisons were to be made between regions that were maximally separated ($A - P, M - L, D - V$), but later it was decided to make all possible comparisons. Figure 2.13 is a graphical representation of some selected arrays of this experiment.

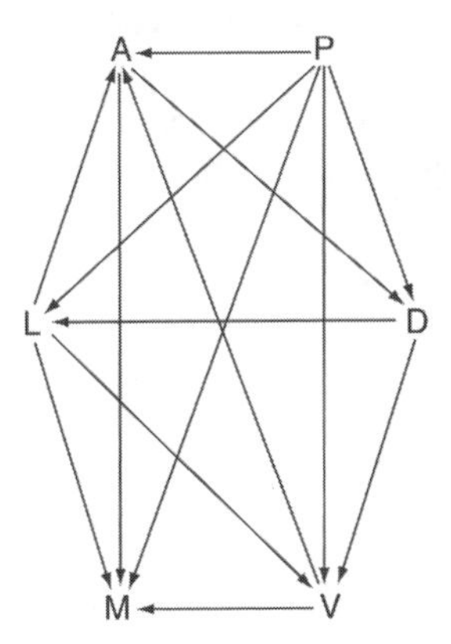

Figure 2.13 *Case study III: The olfactory bulb experiment provided by Dave Lin from the Ngai Lab at the University of California, Berkeley.*

2.4.2 Design

Let us begin our discussion on design with an example comparing three types of mRNA samples ($K = 3$) and let us suppose that all pairwise comparisons are of equal interest. An example of this type of experiment would arise in investigating the differences in expression between different regions (T_0, T_1, T_2) of the brain. The scientific aim of this experiment would be to identify genes that are differentially expressed between different regions of the brain; that is, to identify genes with differential expression between T_0 and T_1, or between T_1 and T_2, or between T_2 and T_0. Figure 2.14 depicts two designs for such a three-level, single-factor experiment, where R denotes a common reference source of mRNA.

For comparing the efficiency of different designs, we fit a linear model and examine the variance associated with the least squares estimates for the parameters of interest. For example, let us consider the design depicted in Figure 2.14b. For any particular gene, we denote the means of the log base 2 intensities across slides by $\alpha_0 = \mathbb{E}\log_2 T_0, \alpha_1 = \mathbb{E}\log_2 T_1$ and $\alpha_2 = \mathbb{E}\log_2 T_2$. As all measurements are paired comparisons, only the differences between the effects are estimable and the contrasts of interest are thus the pairwise differences. For estimation purposes, we can treat T_0 as a "pseudo" common reference. It follows that our parameters of interest are $\phi_1 = \alpha_1 - \alpha_0$ and $\phi_2 = \alpha_2 - \alpha_0$. In another context, we might be more interested in $\alpha_2 - \alpha_1$, and so wish to make that one of our parameters.

We fit the following linear model $y = X\phi + \epsilon$ to the vector y of log-ratios $y_1 = \log_2(T_1/T_0)$, $y_2 = \log_2(T_2/T_1')$ and $y_3 = \log_2(T_0'/T_2')$ from different slides:

$$X = \begin{pmatrix} 1 & 0 \\ -1 & 1 \\ 0 & -1 \end{pmatrix}; \ \phi = \begin{pmatrix} \phi_1 \\ \phi_2 \end{pmatrix} \text{ and } \Sigma = cov(y) = \sigma^2 I.$$

We have used primes (e.g., R') to denote technical replicate material, as this would almost always be the case in experiments like these. Nevertheless, we will begin by assuming that the different log-ratios are independent. The least squares estimates of the parameters are $(X'X)^{-1}X'y$ and the corresponding variances of estimates are given by the diagonal elements of the matrix $\sigma^2(X'X)^{-1}$. Table 2.1 provides comparisons for a few design choices, where for presentation, σ^2 is set to 1. The

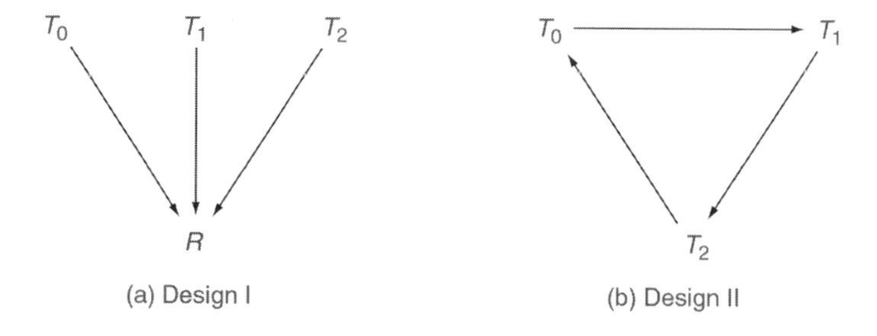

Figure 2.14 *Two designs for a 3-level, single-factor experiment.*

Table 2.1 *Single factor experiment — variance of estimated effects for the three different designs.*[a]

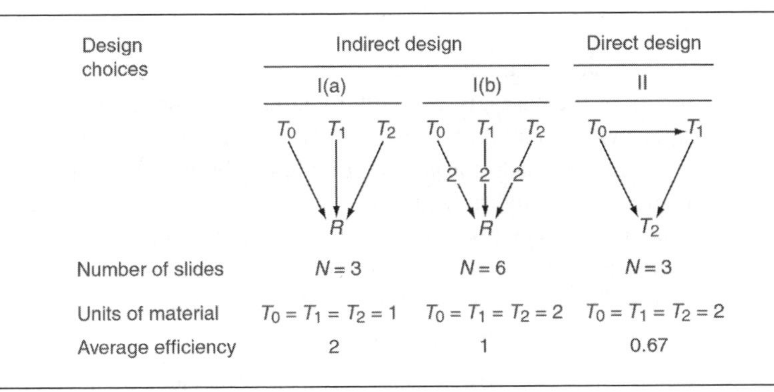

Design choices	Indirect design		Direct design
	I(a)	I(b)	II
Number of slides	$N = 3$	$N = 6$	$N = 3$
Units of material	$T_0 = T_1 = T_2 = 1$	$T_0 = T_1 = T_2 = 2$	$T_0 = T_1 = T_2 = 2$
Average efficiency	2	1	0.67

[a]For presentation, σ^2 was set to 1.

value in each cell of the table is the average variance associated with the three pairwise comparisons of interest: $var(\hat{\phi}_2) + var(\hat{\phi}_1) + var(\hat{\phi}_2 - \hat{\phi}_1)$. This is the same as the A-optimality criterion defined in Kerr and Churchill (2001). This criterion is suitable here because all pairwise comparisons are of equal interest and the main scientific constraint is that they are estimable.

Different design choices will be made depending on different physical constraints. For example, if the experimenters have unlimited amounts of reference material, but only one sample of mRNA from each of T_0, T_1 and T_2, then Design I(a) is the only possible choice out of these three. However, if the experimenters have two samples of mRNA from each of the T_0, T_1, and T_2 regions, then both Designs I(b) and II are feasible, with one using twice as many slides as the other. Direct comparison, Design II, will lead to more precise comparisons between the different regions, and in addition, save on the number of slides required.

In general, extending the designs in Table 2.1 to include K target samples, we considered the following two classes of designs:

(i) *Indirect design.* Perform each $R \rightarrow T_k$ hybridization as well as the corresponding dye-swaps, $T_k \rightarrow R$, $k = 0, \ldots, K-1$, for a total of $2K$ hybridizations.

(ii) *Direct design.* Perform each $T_k \rightarrow T_l$ hybridization, $k, l = 0, \ldots, K-1, k \neq l$, for a total of $K(K-1)$ hybridizations.

For direct designs, the comparison of two samples T_k and T_l can be done *within slides*, while for indirect designs, the contrast $\alpha_k - \alpha_l$ are not estimated directly, but instead *across slides* through the common reference R. It can be shown, using the linear model that estimates of contrasts $\alpha_k - \alpha_l$ have variance σ^2 for indirect designs and variance σ^2/K for direct designs. Adjusting for the different number of slides, the relative efficiency ratio between indirect vs. direct designs (i.e., ratio of variances for estimates of $\alpha_k - \alpha_l$ using both designs) is

$$\frac{v_{(i)}}{v_{(ii)}} = \frac{2K}{K-1}. \tag{2.1}$$

Therefore, for an equivalent number of slides, the limiting efficiency ratio, as the number of treatments K increases, is 2, clearly illustrating the point that direct comparisons within slides yield more precise estimates than indirect ones between slides. Whenever comparisons are made through a common reference R, which alone is not of primary interest, every hybridization involving R is in some sense "wasted," roughly doubling the required number of hybridizations for a given level of precision. As observed before, the relative efficiency is largest for $K = 2$ treatments: where using direct comparisons gives a variance of $1/4$ of that obtainable from an indirect design involving the equivalent number of slides.

Moreover, to compare the two design classes for an equivalent number of target samples, the relative efficiency of the indirect versus the direct design for estimating $\alpha_k - \alpha_l$ is

$$\frac{v_{(i)}}{v_{(ii)}} = \frac{K}{K-1}. \tag{2.2}$$

As the number of treatments K increases, the limiting efficiency ratio is 1, and the number of slides used in the indirect design will be twice as large as that used with the direct design. Comparing this result with Equation 2.1, we see that the differences in efficiency between the two designs are due largely to the amount of material involved in the experiments.

Considering Table 2.1, it is evident that, when K becomes larger, the situation becomes more complex. The analogues of Designs I(a) and I(b) are clear, they are the so-called *reference* designs. The analogue of Design II — which we call *all-pairs* design — is unlikely to be feasible or desirable for a large number of comparisons. For example, with 6 sources of mRNA, there are 15 pairwise comparisons requiring 5 units of each target mRNA; for 7 there are 21, requiring 6 units, and so on. Alternative classes of designs that involve far fewer slides include the *loop* designs of Kerr and Churchill (2001), but these designs can suffer from having long path lengths between some of the comparisons. In Figure 2.15, we offer two alternative designs for six sources of target mRNA and six hybridizations. For Design I, the least squares estimates of contrasts $\alpha_k - \alpha_l$ have variance $2\sigma^2$ for any pairwise comparisons, and hence an average variance of $2\sigma^2$. In contrast, the average efficiency for all pairwise least squares estimates of contrasts in Design II is $1.8\sigma^2$, with some contrasts being relatively more precise (e.g., $var(\hat{\alpha_1} - \hat{\alpha_2}) = \sigma^2$) and others that are relatively less precise (e.g., $var(\hat{\alpha_1} - \hat{\alpha_4}) = 3\sigma^2$). It should be clear from this example that, instead of regarding the problem of choosing a design as a decision between classes of designs (reference, loop, all-pairs), a more productive approach is to ask which comparisons are of greatest interest and which are of lesser interest, and seek a design that gives higher precision to the former and lower precision to the latter. Such a design then will involve a mix of direct and indirect comparisons, tailored to the needs and constraints of the particular context, including the robustness discussed in Section 2.2.7.

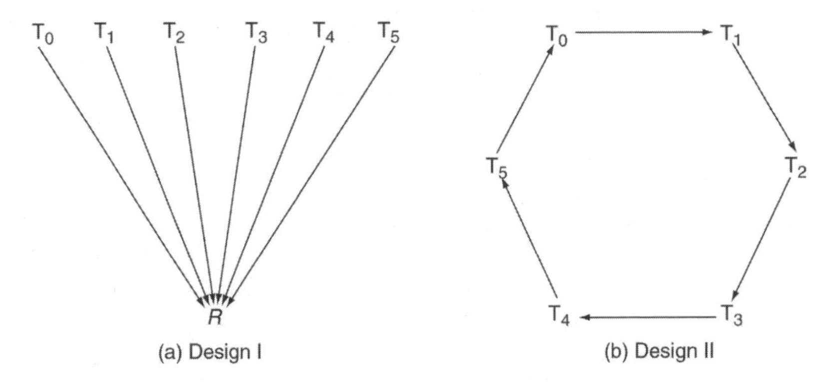

Figure 2.15 *Two possible designs comparing the gene expression between six target samples* $T_0, \ldots, T_5$ *of cells. (a) Indirect comparison: this design compares the expression levels of samples* T_k *and* T_l *to the reference sample separately on two different slides, and estimates the log-ratio* $\log_2(T_k/T_l)$ *by the difference* $\log_2(T_k/R) - \log_2(T_l/R')$*, where samples* R *and* R' *are reference samples. (b) Direct comparison: loop design. This design compares the gene's expression between certain samples* T_k *and* T_l *where* $k, l = 0, \ldots, 5; k \neq l$ *directly on the same slide.*

A more general covariance model

The preceding comparisons do not take into account the different types of replication. Now, let us consider how the efficiencies of our designs change when the covariances arising from the use of technical replicates are included in our analysis. We revisit the designs for a three-level, single-factor experiment depicted in Table 2.1 using Design II first. As before, we denote by $y = (y_1, y_2, y_3)$ the log-ratios observed from slides 1, 2, and 3. To estimate the parameters of interest $\beta = (\phi_1, \phi_2)$ where $\phi_1 = \alpha_1 - \alpha_0$ and $\phi_2 = \alpha_2 - \alpha_0$, we once again fit the linear model by least squares, but this time using generalized least squares. The covariance matrix of the observations is given by:

$$cov \begin{pmatrix} y_1 \\ y_2 \\ y_3 \end{pmatrix} = \sigma^2 \begin{pmatrix} 1 & -\rho & -\rho \\ -\rho & 1 & -\rho \\ -\rho & -\rho & 1 \end{pmatrix}.$$

Here, σ^2 is the variance of a single log-ratio, which in the notation of Section 2.3.9 is $2(\tau^2 - \gamma_1)$, while the covariance between two log-ratios involving a common technical replicate labeled with the same dye in the opposite position is $-\sigma^2\rho = -(\gamma_2 - \gamma_3)$. Analogous linear models and covariance matrices are used for the indirect Designs I(a) and I(b) of Table 2.1. In these cases, we not only have terms in the covariance matrix corresponding to log-ratios sharing a single technical replicate, in Design I(b) we also have terms $2\sigma^2\rho$ corresponding to the covariances between log-ratios sharing two technical replicates. We omit the straightforward details.

With these covariances we can calculate the average variances of the generalized least-squares estimates of the contrasts of interest under the more general model. Because the designs are symmetric, these will be the same functions of ρ as the variance

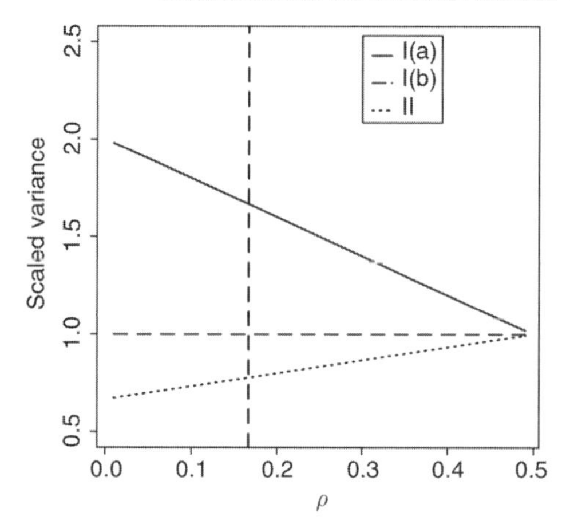

Figure 2.16 *Scaled variances of generalized least squares estimates of the contrast* ϕ_1 *as a function of* ρ *for the three experimental designs I(a), I(b) and II depicted in Table 2.1. The vertical dotted line denotes the estimated* ρ *from the second swirl experiment described in Section 2.3.1.*

$var(\hat{\phi}_1)$ of the generalized least squares estimator $\hat{\phi}_1$ of ϕ_1. Table 2.1 has the results in the case where $\rho = 0$ (i.e. there is zero covariance between log-ratios derived from technical replicate samples). Under the more general covariance model, the variance of the indirect estimate of $\hat{\phi}_1$ from Design I(b) of Table 2.1 is just σ^2, independent of ρ. By contrast, the estimator from Design II, which is a mix of direct and indirect information, increases from $\frac{2}{3}$ to 1 as ρ increases from 0 to 0.5. In fact, in this case the generalized least squares estimator of ϕ_1 coincides with its ordinary least squares estimator. We conclude that the variance of the indirect estimate (Design I(b)) is 1.5 times that of the mix of direct and indirect (Design II) at $\rho = 0$, but comes to equal it at the other extreme $\rho = 0.5$. These observations and the results for Design I(a) can be found in Figure 2.16.

2.4.3 Linear model analyses

General remarks

The approach we adopt to combining gene expression data from replicate experiments to estimate gene expression differences between two types of samples was introduced informally in Section 2.3.3 and Section 2.4.1. In its simplest form, the combination is carried out by averaging observations from individual experiments ("treatment" and "control") and taking differences, to produce a combined estimate of the gene expression effect that is a result of the treatment. This straightforward idea may be extended to experiments such as Design II in Figure 2.14b by employing fixed effects linear models, where we estimate certain quantities of interest (e.g., the $T_1 - T_0$ difference) for each gene on our slide. Further, the idea applies much more generally, and will serve our purpose here, to provide least squares estimates of all estimable contrasts such as the anterior-lateral difference in the bulb experiment of Figure 2.13. As with all our earlier work, we focus on parameters that are, in effect, expectations of log-ratios (e.g., of $\log(A/L)$).

Some small but important differences occur between our use of linear models here and those published in the microarray literature. Others, most notably Kerr et al. (2000), use linear models and the analysis of variance (ANOVA) to estimate expression differences and assess the variability of their estimates. As explained in Section 2.3.4 these writers typically assume a fixed effects linear model for the logged intensities, with terms for dye, slide, treatment, and gene main effects, as well as selected interactions between these factors, and error terms having common variance across genes. Differentially expressed genes are those that exhibit significant treatment $\times$ gene interaction, while normalization is effected by the inclusion of dye terms in the linear model. By contrast, our linear models are for log-ratios from experiments in which normalization has been carried out separately for each slide, typically using a decidedly nonlinear adjustment which could not be captured in a linear model. Indeed, artifacts removed by our nonlinear normalization frequently correspond to interactions assumed absent in the linear approach. We do not include different genes in the same linear model, and so we do not assume a common error variance, and the only terms we do include in the model relate to the mRNA samples and their treatments. We find that our approach deals more satisfactorily with dye differences that are nonlinearly intensity-dependent and spatially dependent in varying ways across a set of slides. Also, our designs are usually not orthogonal, and so would not lead to a unique analysis of variance.

In more recent research, Jin et al. (2001) and Wolfinger et al. (2001) use linear models on a gene by gene basis, with a separate error for each gene, but still including normalization as part of their linear modelling. Their linear models for normalized data also include random effects for arrays. In a sense, these studies are quite ambitious, as the authors took the bold step of treating the signals from the two channels of their cDNA experiment as two separate sources of data, not passing to ratios or log-ratios, but keeping both for their analysis. They have some effects estimated within hybridizations (age), which are therefore based on ratios within hybridizations, and others (sex and strain) that are not, but result from comparisons across hybridizations not expressible as functions of within-hybridization ratios. In our view, this approach can only be adopted following a very thorough multi-slide normalization of all of the single channels, because many systematic nonadditive spatial and intensity-dependent hybridization biases disappear when ratios are taken, but can remain otherwise. When this is done, it would appear that the appropriate analysis should be a two-stratum one, distinguishing the within-hybridization and the between-hybridization strata, for in our experience, the variances in these two strata are likely to be quite different. In brief, the appropriate analysis in this two-stratum context is likely to be significantly more complex than the present one, and for this reason we do not pursue it further here.

Application to case study III

To define the notation used in our linear model, for a typical gene i, let us denote the gene's observed intensity value corresponding to the six different regions of the bulb by A_i, P_i, D_i, V_i, M_i, and L_i. Further, define the average values across hybridizations of the log transformation of these quantities to be $a_i = \mathbb{E}\log_2 A_i$, $p_i = \mathbb{E}\log_2 P_i$, $d_i = \mathbb{E}\log_2 D_i$, $v_i = \mathbb{E}\log_2 V_i$, $m_i = \mathbb{E}\log_2 M_i$, and $l_i = \mathbb{E}\log_2 L_i$. To estimate the spatial gene expression for gene i, we fit the following linear model:

$$y_i = X\beta_i + \epsilon_i,$$

where y_i is the vector of log-ratios from all the slides; X is the design matrix; ϵ_i is a vector of the disturbance term; and β_i is a vector of parameters. The five estimable parameters we choose are given by $a_i^{(l)} = a_i - l_i$, $p_i^{(l)} = p_i - l_i$, $d_i^{(l)} = d_i - l_i$, $v_i^{(l)} = v_i - l_i$ and $m_i^{(l)} = m_i - l_i$. In addition, we assume that the error terms associated with the different slides are independent and identically distributed with $\mathbb{E}(\epsilon_i) = 0$ and $cov(\epsilon_i) = \sigma_i^2 I$, where I is the identity matrix. Note that in this experiment, some but relatively little use was made of technical replicates. For simplicity, we ignore possible correlations in our discussion. The design matrix and parameters corresponding to the data of case study III is given next.

$$\begin{pmatrix} y_{1i} \\ y_{2i} \\ y_{3i} \\ y_{4i} \\ y_{5i} \\ y_{6i} \\ y_{7i} \\ y_{8i} \\ y_{9i} \\ y_{10i} \\ y_{11i} \\ y_{12i} \\ y_{13i} \\ y_{14i} \\ y_{15i} \end{pmatrix} = \begin{bmatrix} 1 & 0 & 0 & 0 & 0 \\ 0 & -1 & 0 & 0 & 1 \\ -1 & 0 & 0 & 0 & 1 \\ 0 & 0 & 0 & 0 & 1 \\ -1 & 1 & 0 & 0 & 0 \\ 0 & 0 & -1 & 1 & 0 \\ -1 & 0 & 1 & 0 & 0 \\ 0 & -1 & 1 & 0 & 0 \\ 0 & -1 & 0 & 1 & 0 \\ 0 & 0 & 0 & 1 & 0 \\ 1 & 0 & 0 & -1 & 0 \\ 0 & -1 & 0 & 0 & 0 \\ 0 & 0 & -1 & 0 & 0 \\ 0 & 0 & 0 & -1 & 1 \\ 0 & 0 & 1 & 0 & -1 \end{bmatrix} \cdot \begin{pmatrix} a_i^{(l)} \\ p_i^{(l)} \\ d_i^{(l)} \\ v_i^{(l)} \\ m_i^{(l)} \end{pmatrix} + \epsilon_i. \qquad (2.3)$$

The parameter β_i can be estimated by $\ddot{\beta}_i = (X'X)^{-1}X'y_i$. In practice, we used a robust linear model, so that our estimates are less affected by outliers. The data are fitted to the linear model described previously by iteratively re-weighted least squares (IWLS) procedure using the function `rlm` provided in the library MASS in the statistical software package R. Details of the theory and implementation of the "robust linear model" can be found in Venables and Ripley (1999).

Although we did not carry out the bulb study with a common reference mRNA source, this was initially contemplated, and we would have used "whole bulb" mRNA, W. Had we done so, we would have had seven mRNA samples and could then have estimated six parameters. The natural parameterization would then have been $a_i^{(w)} = a_i - w_i$, $p_i^{(w)} = p_i - w_i$, $d_i^{(w)} = d_i - w_i$, $v_i^{(w)} = v_i - w_i$, $m_i^{(w)} = m_i - w_i$, and $l_i^{(w)} = l_i - w_i$, where $w_i = \mathbb{E} \log_2 W_i$, and the analysis would have begun with robustly fitting a suitable linear model to estimate these parameters.

Returning to the parametrization we did use, what came next? It is clear from the design matrix that all 15 pairwise comparisons of the form $a_i - l_i$ are estimable, and so our first summary description of a gene's spatial pattern was the profile consisting of the set of 15 parameter estimates of this form. Not surprisingly, profiles of this kind were not as easy to interpret as we would have liked, having a high degree of redundancy. Accordingly, we switched to the more economical profile consisting of the estimates of the six parameters a_i, p_i, d_i, v_i, m_i and l_i, subject to the zero sum constraint $a_i + p_i + d_i + v_i + m_i + l_i = 0$, that is, to estimates of the six parameters $a_i - \frac{1}{6}(a_i + p_i + d_i + v_i + m_i + l_i), \ldots$. These profiles were much easier to visualize, and so we worked with them in the subsequent analysis. We refer to Lin et al. (2002) for the rest of the story, noting here that our main focus following the preceding linear model analysis was the clustering of the spatial profiles. We restricted ourselves to profiles which seemed to be real, that is, not noise, and we used Mahalanobis distance between profiles based on (robust variant of) the variance-covariance matrix $\sigma^2(X'X)^{-1}$ of the parameter estimates.

Remarks on assessing parameter estimates

Most of the issues we discussed in Section 2.3 have analogues in this more general setting involving linear models for expectations of log-ratios. This includes the use of estimated standard deviations for estimates of parameters in the linear models, t-statistics (which are standardized parameter estimates here), modified t-statistics, robust estimates, empirical Bayes statistics, and the use of Q–Q plots and more formal tests of significance. Once we leave replicated treatment vs. control or multiple-treatment experiments, few opportunities exist for permutation-based, distribution-free methods, but the bootstrap appears to be a promising alternative. The techniques just mentioned are still relatively unexplored in this context, no doubt because the use of linear models with microarray data has so far been limited. As the use of linear models increases, we can expect all the analogues of techniques effective in treatment vs. control experiments to find their place in this more general setting.

2.4.4 Time-course experiments

This class of microarray experiments is among the most widely analyzed and reanalysed in the literature, with a relatively large number of papers discussing one or more of the cell-cycle datasets (Cho et al., 1998; Chu et al., 1998; Spellman et al., 1998). We will not attempt to review all this literature, but instead comment briefly on the general approaches and give references. A major drawback of this methods-based summary is the absence of any careful discussion of the biological motivation of the methods, present in some but by no means all of the articles we mention.

Microarray time-course data differs from traditional time series data in that they tend to be either short and irregularly spaced (e.g., 0, 1, 4, 12, and 24 hours after some intervention), or longer and equally spaced, but still with relatively few time points, typically only 10 to 20. The cell cycle data involved about 16 time points and included just two full cycles, which is quite short by normal time series analysis standards. A further common feature of such data (e.g., developmental time-courses) is that there are few if any *a priori* expectations concerning the patterns of responses of genes, apart from vague notions such as "may come on some time, stay on for a while, and then go off," or will be on "early" or "late" in the time period observed. Depending on the time increments and the context, smoothness of a gene's temporal response may or may not be a reasonable expectation. All this suggests that the analysis of microarray time-course data will offer many new challenges to statisticians, and that it will frequently be more exploratory than model-based.

Bearing these remarks in mind, it is not surprising that initial approaches to analyzing time-course data relied on simple mathematical modeling and cluster analysis (see Chu et al., 1998; Cho et al., 1998; Spellman et al., 1998; and Chapter 4 of this book). Indeed, clustering is still the most widely used technique with time-course data, although regression and other model-based statistical methods are becoming more common.

Design

A short time-course experiment may be viewed as a single factor experiment with *time* being the factor. The additional information in this context is that there is a natural ordering between the different target samples, in contrast with the experiment described in Section 2.4, where no ordering existed between the different samples. The ordering of the "levels" of the factor time will single out certain comparisons (e.g., each time with baseline, or between consecutive time points) and linear contrasts (e.g., concerning linearity, concavity, or monotonicity) of interest to the researcher, and so the design choice will definitely depend on the comparisons of particular interest. Further, the best design can depend critically on the number of time points.

In a small study (e.g., with four time points), finding the design that is optimal in some suitable sense can be done simply by enumerating all possibilities (see Table 2 in Yang and Speed (2002) for some simple examples). It is not feasible to enumerate all possible designs for problems with a much larger number of time points, such as the cell-cycle experiment. Developing algorithms for finding near-optimal designs is a research topic of interest here.

Clustering time-course data

Noteworthy early research on large-scale gene expression was conducted by Wen et al. (1998), an RT-PCR study of 112 genes, each measured on cervical spinal cord tissue in triplicate at nine different time points during the development of the rat central nervous system. These authors clustered the temporal responses of the genes, and found five main groups. Their approach is typical of that many other authors have taken since then with microarray data. Tamayo et al. (1999) introduced self-organizing maps into the microarray literature in the context of time-course data, illustrating them on the yeast cell-cycle data and another dataset concerning hematopoietic differentiation in four activated cell lines, each sampled at 4–5 times. Other research adopting a similar approach include Saban et al. (2001) and Burton et al. (2002), both of which have just 5 time points, with 4 and 2 replicates, respectively, at each time. Langmead et al. (2002) provides a critique of the Fourier approach to short time-course data, and offers a novel analysis incorporating autocorrelation and a new metric for clustering. Their methods were applied to simulated data and the cell-cycle datasets.

Principal component, singular value decompositions, and related methods

It is worth separating this category of methods from the clustering, although in most cases the aims are very similar. Chu et al. (1998) and Raychaudhuri et al. (2000) are early examples of the application of principal component analysis (PCA) to microarray time series data. Alter et al. (2000) address similar issues using singular value decompositions (SVD), but go much further in creatively displaying the results. Ghosh (2002) uses resampling methods to estimate the variability in SVD, and applies his methods to the Cho et al. (1998) data. A dynamic model of gene expression is presented in Holter et al. (2001), who estimate the time-independent translation matrix in a difference equation model for the "characteristic modes" (i.e., the eigenvectors of the SVD of published time-course datasets), (Spellman et al., 1998; Chu et al., 1998; Iyer et al., 1999), using 12, 7, and 13 time points, respectively.

Regression and related model-based approaches

In a series of articles, (Zhao et al., 2001; Thomas et al., 2001; Xu et al., 2002), L. P. Zhao and colleagues promote the use of appropriately defined regession models to analyze microarray data. The first and third of these articles discuss time-course data (Zhao et al., 2001) and revisit the cell-cycle data. Perhaps the most interesting of these papers is Xu et al. (2002), which analyzes parallel eight-time-point series from both transgenic and control mice. This article presents an approach to the joint analysis of related series which can be compared with the one discussed briefly in Section 2.5.2.

Two articles that make very explicit use of the temporal pattern expected of (or sought from) genes are Kato et al. (2000) and Sasik et al. (2002). Both invoke differential equation models for mRNA concentrations, and solve them under special assumptions to arrive at regression equations, which are then fitted to the data. Kato et al. (2000) is the more ambitious of the two, attempting to infer genetic networks from the data using multiple regression. They apply their method to an early yeast dataset from Derisi et al. (1997) involving seven different times. By contrast, Sasik et al. (2002) model genes in isolation. Their data were sampled every 2 hours for 24 hours during the developmental program of the slime mold *Dictyostelium discoideum*. These authors begin their analysis by fitting to the data on each gene a simple nonlinear kinetic model based on a first-order differential equation. The equation embodies a threshold model for transcription, and focuses on genes which come on "sharply." They then assess the fit of this model to each gene, and go on to examine the temporal patterns of the genes which fit the model satisfactorily.

Another noteworthy article essentially adopting a regression approach is Shedden and Cooper (2002), which presents a critical analysis of the Cho et al. (1998) cell-cycle data. In fact, their reanalysis suggests that "all apparent cyclicities in the expression measurements may arise from chance fluctuations," and that "there is an uncontrollable source of experimental variation that is stronger than the innate variation of gene expression in cells over time." In light of the large number of reanalyses of the cell-cycle data, this article is well worth reading.

The use of Bayesian networks is most appropriately noted under this heading (Friedman et al., 2000). This is an ambitious attempt to infer gene interactions from the cell-cycle data of Spellman et al. (1998), one in which the role of time in the analysis is by no means straightforward.

Aach and Church (2001) show how to align two time series when the time scales might have become "warped" due to different time sampling procedures or the different rates at which common biological processes evolve in related experiments. This notion is also present in Bar-Joseph et al. (2002), who present a rather general model-based approach to clustering, supposing that each gene's time-series can be viewed as having gene-specific and class-specific parameters multiplying spline basis functions, with Gaussian noise added. The estimation is by maximum likelihood and a byproduct of this is an assignment of genes to classes. Both of these articles illustrate their methods on the yeast cell-cycle data of Spellman et al. (1998), while Bar-Joseph et al. (2002) also analyze a later yeast dataset (Zhu et al., 2000).

Articles by Butte et al. (2002) and Filkov et al. (2002) focus on analytical issues underlying the identification of pairs of genes that are co-regulated in time-course experiments, and both illustrate their approach on the cell cycle datasets Cho et al. (1998) and Spellman et al. (1998). Our final reference in this category is to Klevecz (2000) and Klevecz and Dowse (2000) who also study the cell cycle data, this time using wavelets. Their interesting analyses suggest some quite novel conclusions, including the possible regulation of the yeast cell cycle by "an attractor whose fundamental period is an emergent property of dynamic interactions within the yeast transcriptome."

Contrasts

A simple and potentially powerful approach to extracting information from time-course data makes use of linear combinations (or functionals) of the expression values across times. Following standard statistical usage, we will call these *contrasts*, which usually but do not always require that the coefficients sum to zero. The coefficients in contrasts can be tailored to the aims of the analysis, and they can be fixed *a priori*, or determined by the data. An example of a data-driven contrast would be one defined by the expression profile of a prespecified gene, something we might use in seeking other genes with temporal patterns of expression match those of that gene. A fixed set of coefficients might be one that estimates a linear trend. Another example is the discrete Fourier transform, which can be viewed as a fixed set of constrasts (corresponding to different frequencies) evaluated on an input series. Contrasts thus generalize many familiar calculations, and as we shall see shortly, they also generalize single variable regression modeling. When we have more than one contrast, it is not uncommon to orthogonalize them sequentially, in the hope that, with approximately normally distributed input data, the resulting estimated contrasts are approximately independent. This is not necessary, however.

When $c_1, c_2, \ldots$ is a set of contrast coefficients corresponding to times $1, 2, \ldots,$ and $E_1, E_2, \ldots$ are the corresponding expression time series for a given gene, the sum $< c, E > = \Sigma_t c_t E_t$ is the value of the contrast for that gene. If we normalized this quantity by $|c|^2 = \Sigma_t {c^2}_t$, supposing for simplicity that $\Sigma_t c_t = 0$, then $< c, E > /|c|^2$ would be the regression coefficient corresponding to the fit of expression values to the temporal pattern (c_t). As this normalizing quantity is the same for all genes, there is no real need to include it in the fitting and assessing procedure. We note here that the expression values E_t may be "absolute," as with Affymetrix chip data, or "relative" as with two-color cDNA or long oligonucleotide data, and these will usually be on the log scale.

Coefficients have many natural candidates (c_t), and we simply illustrate with a few that we have found helpful. In a time-course experiment (Lönnstedt et al., 2002), where mRNA from a cell line was sampled at 0.5, 1, 4, and 24 hours following stimulation with a growth factor, we sought "early" and "late" responding genes. An *ad hoc* but apparently useful working definition of these genes was as follows: those genes with large values of $< c, E >$ with $c_t = (t - 24.5)^2$ were termed early, and those with $c_t = t^2$ (i.e., with $c = (0.25, 1, 16, 576)$) were termed late, respectively. It was an easy matter to obtain candidates for these genes: we simply did a Q–Q plot of the values $< c, E >$ and somewhat arbitrarily selected cutoffs determining genes with unusually high or low values. More formal tests of significance are clearly possible, just as they were with similar contrasts defining gene expression differences.

The theme of contrasts in the analysis of time-course data is prominent in Fleury et al. (2002b), Fleury et al. (2002a), and Hero and Fleury (2002). In these articles, many examples of the use of contrasts are given, and the procedure for selecting genes from the various *Pareto fronts* is illustrated (refer to Section 2.3.6). In essence, Pareto fronts and the variants presented in these articles all seek to identify genes that have large values for *all* of a set of contrasts of interest. This ends our brief survey of models, methods, and the literature on the analysis of time-course experiments.

2.5 Factorial experiments

The previous examples have all been single-factor or one-way designs, where the factor has two, three, or more levels. A more complex class of designs arises when two or more factors are considered jointly, each factor having two or more levels. These *factorial experiments* are used to study both the expression differences caused by single factors alone, as well as those resulting from the joint effect of two or more factors, especially when it differs from what might be predicted in the basis of the factors separately, the phenomenon known as *interaction*. Factorial experiments were introduced by R.A. Fisher in 1926, and studied extensively by his collaborator Yates (1937). More recent references are Cox (1958) and Box et al. (1978).

2.5.1 Case study IV: the weaver mouse mutant

This is a case study examining the development of certain neurons in wild-type (wt) and weaver mutant (wv) mice, Diaz et al. (2002a). The weaver mutation affects cerebellar granule neurons, the most numerous cell-type in the central nervous system. In the near-absence of these cells, the mice have a weaving gait. The wv mutant mouse strains were purchased from the Jackson Laboratory, and their genotypes were determined by a restriction site-generating PCR protocol. In the mutant mice, granule cells proliferate in the external granule cell layer, but terminally differentiated cells die before they migrate to the internal granule cell layer. As a result, the weaver mutants have greatly reduced numbers of cells in the internal granule cell layer, in comparison with the wt mice of the same strain. Consequently, the expression of genes which are specific to mature granule cells or expressed in response to granule cell-derived signals is greatly reduced. Figure 2.17 is a graphical representation of four selected slides from Diaz et al. (2002a) in a form convenient to illustrate the parametrization of factorial experiments that we will use.

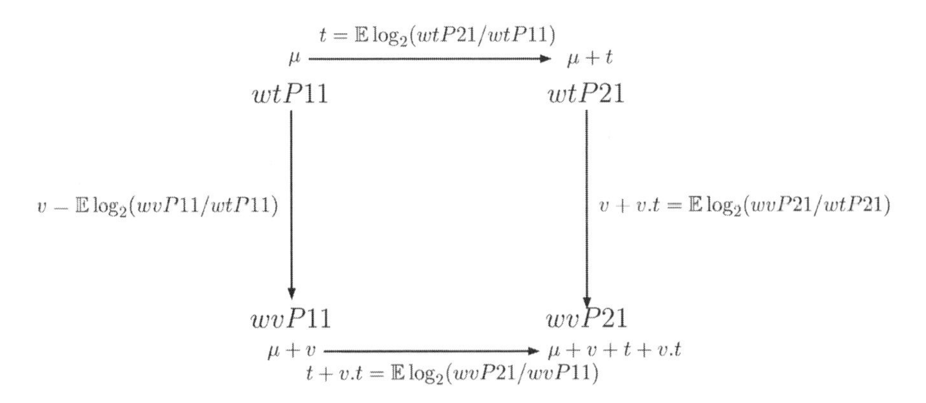

Figure 2.17 *Case study IV: A portion of the weaver experiment provided by Elva Diaz from the Ngai Lab at the University of California, Berkeley. One parametrization for this* 2×2 *factorial experiment is indicated, see text for more explanation.*

2.5.2 *Design and analysis of factorial experiments*

In this 2×2 factorial experiment, gene expressions levels are compared between the two strains wt and wv at the two postnatal times $P11$ and $P21$ (days). The four possible mRNA samples are $wtP11$, *wild-type* at postnatal day 11, $wvP11$, *weaver* at postnatal day 11, $wtP21$, *wild-type* at postnatal day 21, and $wvP21$, *weaver* at postnatal day 21.

Figure 2.17 shows one parametrization for this experiment. Here, μ is the expectation of the log intensity $\log wtP11$, $\mu + t$ is the expectation of $\log wtP21$, etc. for a generic gene. Note that our parametrization is different from the symmetric ANOVA parametrization used in most statistics text books. We do it this way because we find the terms are more readily interpretable to biologists. Thus, the main *mutant* effect v is the expectation $\mathbb{E}\log_2(wvP11/wtP11)$ of the gene expression difference (on the log scale) between wv and wt at P11. Similarly, the main *time* effect $t = \mathbb{E}\log_2(wtP21/wtP11)$ is the expected gene expression difference (on the log scale) between days 21 and 11 for wt mouse. The *mutant* by *time interaction* is defined by

$$v.t = \mathbb{E}\log_2 \frac{wvP21/wvP11}{wtP21/wtP11}.$$

Genes with a nonzero interaction term can be interpreted as genes for which the gene expression difference between $wtP11$ and $wtP21$ is different, on average, from that between $wvP11$ and $wvP21$; equivalently, genes whose expression difference between $wtP11$ and $wvP11$ are different, on average, from those between $wtP21$ and $wvP21$.

How do we design 2×2 factorial experiments in this context? Four possible designs for the weaver mutant case study, each involving six hybridizations, are represented in Table 2.2, where here we follow Glonek and Solomon (2002). The designs head the columns, and the table entries are the corresponding variances of the ordinary least squares estimates of the main effect parameters v and t and the interaction parameter $v.t$. Suppose that our main goal is to identify genes with large interaction with six hybridizations. Designs II and IV give the smallest variance for the interaction term, but the main effect for t is not even estimable in Design IV. Design I, which many biologists would use instinctively (perhaps without the dye-swaps), is by far the worst for precision in estimating the interaction. The preferred design will depend on the level of interest in the main effects in relation to the interaction, assuming all physical constraints are satisfied. In general, Design II (or its complement, with dye-swaps horizontally instead of vertically) will probably be the design of choice, offering good precision for all comparisons, though more for one main effect than the other.

The preceding design options help us to make an important general design point, namely, that in addition to experimental constraints, design decisions will be driven by an awareness of which effects are of greater interest to the investigator, and which are of lesser interest. More fully, the effects for which the greatest precision is required should be estimated *within slides* to the greatest extent possible, while effects of lesser interest can be measured less precisely, *between slides*. In extreme cases, where there

Table 2.2 2×2 *factorial experiment — variance of estimated effects for the four different designs.*[a]

	Indirect design	A balance of direct and indirect		
Design choices	I	II	III	IV
	$wvP11$ $wtP21$ $wvP21$ / $wtP11$	$wtP11$ $wtP21$ / $wvP11$ $P21$	$wtP11$ $wtP21$ / $wvP11$ $wvP21$	$wtP11$ $wtP21$ / $wvP11$ $wvP21$
Main effect t	0.5	0.67	0.5	NA
Main effect v	0.5	0.43	0.5	0.3
Interaction $v.t$	1.5	0.67	1	0.67

[a]For presentation, σ^2 was set to 1.

is no interest at all in quantifying an effect, it need not even be estimable within slides (refer to Design IV in Table 2.2). Similar points were made by Kerr and Churchill (2001), who recommended greater use of loop designs.

We now revisit the design problem just discussed when correlations between technical replicate data are included. Figure 2.18 provides a representation of this 2×2 factorial experiment (Design II in Table 2.2) and the number next to the arrows in the diagram is the slide number. Here, the parameters of interest are the main effect v, main effect t, and the interaction effect $v \cdot t$.

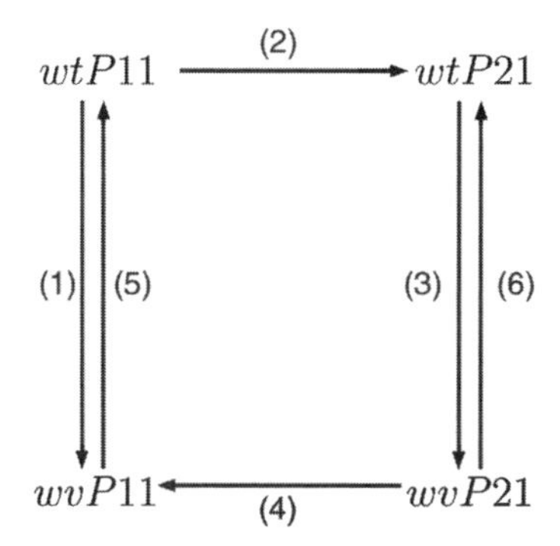

Figure 2.18 *Experimental Design II of Table 2.2, with individual hybridizations numbered.*

For the observed log-ratios $y = (y_1, \ldots, y_6)$, we fit the following linear model:

$$\mathbb{E}\begin{pmatrix} y_1 \\ y_2 \\ y_3 \\ y_4 \\ y_5 \\ y_6 \end{pmatrix} = \begin{pmatrix} 1 & 0 & 0 \\ 0 & 1 & 0 \\ 1 & 0 & 1 \\ 0 & -1 & -1 \\ -1 & 0 & 0 \\ -1 & 0 & -1 \end{pmatrix} \begin{pmatrix} v \\ t \\ v.t \end{pmatrix},$$

and

$$\Sigma = cov\begin{pmatrix} y_1 \\ y_2 \\ y_3 \\ y_4 \\ y_5 \\ y_6 \end{pmatrix} = \sigma^2 \begin{pmatrix} 1 & \rho & 0 & \rho & -2\rho & 0 \\ \rho & 1 & -\rho & 0 & -\rho & \rho \\ 0 & -\rho & 1 & -\rho & 0 & -2\rho \\ \rho & 0 & -\rho & 1 & -\rho & \rho \\ -2\rho & -\rho & 0 & -\rho & 1 & 0 \\ 0 & \rho & -2\rho & \rho & 0 & 1 \end{pmatrix}.$$

Here, as in Section 2.4.2, we denote the covariance between two log-ratios that share a technical replicate term in the same position (numerator or denominator) by $\sigma^2\rho$, with the signs being reversed if the shared term is in the opposite position, while the covariance is $2\sigma^2\rho$ or $-2\sigma^2\rho$ between log-ratios with two technical replicate terms in common. The diagonal elements of the matrix $(X'\Sigma^{-1}X)^{-1}$ provide the estimates for $var(v)$, $var(t)$ and $var(v \cdot t)$, and similarly we can estimate these variances for all four designs shown in Table 2.2.

Figure 2.19 compares the variances of the generalized least squares parameter estimates for different values of ρ. As we have already remarked, the main effect t is not estimable for Design IV. In general, the variances increase as the correlation among technical replicates increases, because we do not gain as much independent

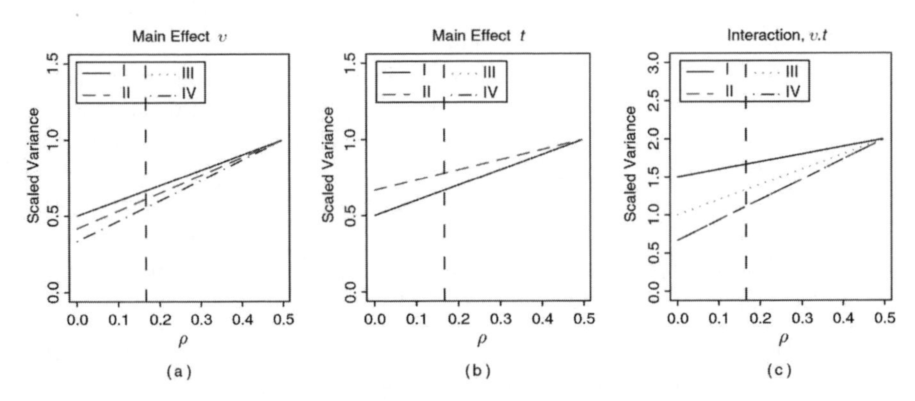

Figure 2.19 *Comparing variances for a series of ρ values for parameter estimates from four designs for a 2×2 factorial experiment (see Table 2.2). Plotted are scaled variances for (a) main mutant effect v; (b) main time effect t; and (c) interaction effect $v.t$, for ρ ranging from 0 to 0.5. The vertical dashed line in all three panels denotes the estimated ρ from the second swirl experiment described in Section 2.3.1.*

information while averaging, although this is not the case with the estimate of the interaction parameter in Design I. Perhaps the most striking conclusion from this figure is that for high enough correlation between technical replicate Design I becomes competitive for estimating the interaction parameter. Note also that the designs which lead to estimates of parameters with equal variance when $\rho = 0$ (refer Table 2.2), specifically Designs I and III for the main effects t and v, Designs II and IV for the interaction parameter, also lead to estimates with equal variances for these parameters for all ρ.

It is worth pointing out that by showing how to take covariances between target samples into account, our aim here is simply to help make more informed design decisions. We do not advocate the use of technical replicates in place of biological replicates. As explained in Section 2.2.4, the type of replication to be used in a given experiment depends on the *precision* and on the *generalizability* of the experimental results sought by the experimenter. In general, an experimenter will want to use biological replicates to validate generalizations of conclusions. Often technical replicates are the result of physical constraints on the amount of mRNA samples an investigator can obtain. For example, in the Case Study III, the investigators were comparing small regions of the brain from new born mice. In these experiments, the material is rare and an amplification technique is used to generate more material. This technique will inevitably introduce correlation between subsamples of the amplified material. Thus, we see that at times the consideration of covariances between technical replicates is a form of physical constraint that we need to recognize in making design decisions. In summary, when making designs choices, investigators should consider the pattern and approximate level of correlation between technical replicates, in addition to identifying which effects are of greater and which are of lesser scientific interest.

2.5.3 Analysis of factorial experiments

We have already discussed the estimation of parameters in linear models for log-ratios (see Section 2.4.3); it all applies to our main effects and interaction parameters here. Figure 2.20 shows a plot of the estimates of the interaction parameter $v \cdot t$ (vertical axis) against the average values $\bar{A}$ (horizontal axis) for the genes involved in Case Study IV. We have retained the coloring convention of Figure 2.6, and we observe that there appears to be a number of genes with quite large interactions by all three of the measures (estimated effect size, standardized estimate of effect size, and the Bayesian compromise between these two). We could conduct other analyses, including Q–Q plots, but we are content with just this one illustrative plot.

Most of what we have discussed for 2×2 factorial experiments extends straightforwardly to 2×3, 2×4, 3×3, and more general two-way factorial experiments, as well as to $2 \times 2 \times 2$ and other higher-way factorial experiments. It should all be quite clear how to proceed, but we need a special comment. We continue to regard nonstandard parametrizations analogous to the one used previously as preferable to the usual ANOVA-type ones from the viewpoint of the biological interpretability of the parameters.

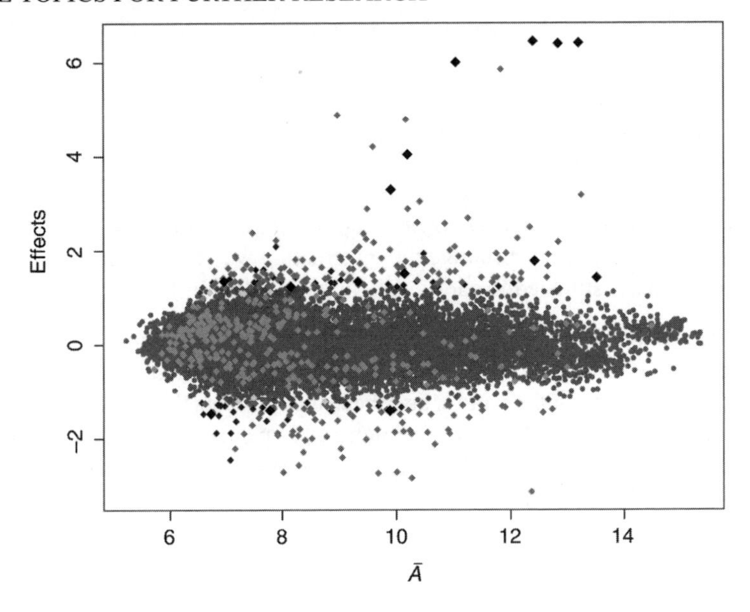

Figure 2.20 *(See color insert following page 114.) Estimated effect $\hat{v.t}$ vs. $\bar{A}$ plot for the interaction $v.t$ between time and mutant.*

2.5.4 Comparison of time series as factorial experiments

One area in which we have made considerable use of factorial experiments is worth discussing further: comparing tissue from two (or more) different mouse strains (e.g., *wt* and *mutant*) over several times (i.e., in extensions of studies such as IV above to more than two times). In the microarray literature, as in many other places, it is common to regard time as a *covariate* instead of a *factor*, as we are suggesting here, and use a more traditional regression approach (see Section 2.4.4). We have found that for a small number of times (e.g., 2 to 5), where there is no special spacing or *a priori* expectation of particular patterns, particularly in developmental studies, it is advantageous to treat time as a factor. When doing so, we find that particular interest attaches to interactions of other factors with time, frequently focusing on just the points of biological interest to the experimenter. The interaction in case study IV is one such example.

2.6 Some topics for further research

A large number of problems occur in the design and analysis of microarray data, on which further research is required. Some problems are implicit in the discussion we presented earlier (e.g., on design, the analysis of comparative and time-course experiments, where the incompleteness of our knowledge is abundantly clear). In this short section, we touch on a few problems that were not discussed elsewhere in this chapter.

2.6.1 *Involving $\bar{A}$ in ranking genes for differential expression*

In Section 2.3.6, we discussed why we think it desirable to take notice of a gene's average intensity value $\bar{A}$ in determining whether it is differentially expressed. At present, none of the measures $\bar{M}$, t, or B that we have described, or the many variants on them, do this in any direct way. To the extent that variability is frequently lower at the higher end of the $\bar{A}$-range, standardizing by SD goes part of the way toward this end, but it does not fully incorporate the information we see in the plots of $\bar{M}$ (or its analogues such as estimated linear model parameters) against $\bar{A}$.

A first step that we have used with some satisfaction is the following simple approach. We block the $\bar{A}$-range into 20 intervals; within each, we calculate upper and lower limits containing 95% of the values, and then smoothly join the values across the full range. Ideally, we would like this for a number of percentages, such as 90%, 95%, 99%, 99.9%, etc., and then one could determine (to some extent) the total number of points outside the range. What appears to be lacking is a formal approach to this question, one that recognizes and deals with the failure of the log transformation to be homoscedastic over the whole range of $\bar{A}$, as well as the changing density of points over this range.

Our last comment raises the question of transformations other than log-ratios of the intensities measuring gene expression. Not surprisingly, a number of statisticians have considered this topic, including Rocke and Durbin (2001), Durbin et al. (2002) and Huber et al. (2002). In our view, it is highly desirable to work with log-ratios if at all possible, as this is a simple and intuitive transformation for biologists, especially when compared with differences of $arcsinh$. Although there considerable heteroscedacity always exists in the distribution of untransformed intensities, it is usually greatly reduced although not completely eliminated after taking logs. The authors just cited use a mean-variance relationship which includes a dominant additive component of variation at low intensities and a dominant multiplicative component at high intensities. One consequence of this is that the transformations they derive behave like log at high intensities, despite a body of evidence that this is too severe a transformation in that range. Further, we believe that the need for an additive component at low intensities can be reduced if not completely removed by better background adjustment, see Yang et al. (2002a). In brief, we are not sure that the problem discussed here can be solved by using a transformation of intensities other than the log-ratio.

2.6.2 *Multiple gene analyses*

It is not hard to see that, for the most part, the approaches to design and analysis offered so far in this chapter can be described as *a single gene approach*. We summarize them as follows:

1. Find appropriate ways of carrying out inference for single genes, and do this with each gene separately.

2. Deal with the special problems that arise (e.g., multiple testing, empirical Bayes) because we have tens of thousands of genes.

In brief, we have outlined a one-gene-at-a-time approach. On the other hand, all the many forms of cluster analysis used with microarray data (and discussed in Chapter 4) are explicitly all (or most or many) genes at a time, and for the most part do not focus on single genes. It is clear to us that room exists for approaches to statistical inference, that is, to answering biologists' questions. These questions are, in a suitable sense, midway between single gene and cluster analyses.

Here is a line of reasoning that leads naturally to multiple gene analyses. Suppose that we seek pairs of genes that are jointly differentially expressed in a treatment-control comparison. In many cases, these genes would be differentially expressed *separately*, and therefore should show up in single gene analyses, but there may well be gene pairs that are significantly affected *jointly*, but not much so separately. For example, the association between two (or more) genes may be different under one condition from that under another. Can we find the gene pairs so affected? One approach we have tried, which failed miserably, was to go beyond the single gene one- or two-sample t-tests outlined in Section 2.3 to one or two-sample bivariate Hotelling T^2-tests. With an experiment similar to case study II having eight treatment and eight control slides, we found nothing useful: the millions of pairs of genes led to T^2-statistics that were hardly distinguishable from noise. It became clear that a "head-on" approach was doomed to failure because of the sheer number of pairs of genes.

The following question logically precedes the search for pairs of genes that are jointly differentially expressed: Can we find pairs (more generally sets) of genes with transcriptional levels that are associated in an experiment? Questions like this have been the subject of a series of articles by S. Kim, E. Dougherty, and colleagues (Kim et al., 2000a; Kim et al., 2000b; Dougherty et al., 2000). In these articles, two-channel microarray data are reduced to ternary form ($-1, 0$, and $+1$ for downregulated, invariant, and upregulated, respectively) and nonlinear systems, such as neural networks, are used to determine when one gene's transcriptional response can be predicted by that of other genes. With thousands of genes, and thus millions of sets of predictors, this is clearly a formidable task, and yet it is but a small beginning down the road of true multiple gene analyses. We refer readers to these articles and commend the general problem to them.

A final example of multiple gene analyses is the synthesis of classification and clustering that leads to a more formal statistical treatment of clustering than is usually the case. Here, we simply refer the reader to Fridlyand and Dudoit (2001) and references therein.

2.6.3 Significance testing

We have already indicated that the theory for multiple testing in the microarray context is not satisfactorily complete, even for simple problems such as identifying differentially expressed genes. A number of other contexts exist that involve microarray data in which biologists would like to make statements concerning statistical significance and where the multiple testing issue arises. Two such contexts are described next.

In the package GenMAPP (`http://www.genmapp.org/about.asp`) genes belonging to a given biochemical pathway can be colored red or green or yellow according to whether they are up- or down-regulated or unchanged, in some treatment mRNA relative to mRNA from a reference condition. If the genes in a pathway are in truth unaffected by the treatment, one would expect the pattern of red and green coloring to be "random" in some suitable sense, whereas if that pathway is indeed affected by the treatment intervention, one might expect more genes changing in one direction or another than might occur "by chance." Can these notions be made rigorous and a test of significance developed for the null hypothesis that a specified biochemical pathway is not affected by treatment intervention? For a first attempt at addressing this question, see Doniger et al. (2002). What about the same question when we do not specify the pathway in advance, but search over pathways after the analysis?

A very similar question arises when biologists make use of the gene ontology (GO) (`http://www.geneontology.org/`), which is a framework for assigning genes a molecular function, a biological process and a cellular location. Naturally, in most cases, the assignments should be regarded simply as principal ones, for many, perhaps most, genes will not have unique assignments to categories within these headings. Nevertheless, the GO is extremely valuable to biologists seeking to interpret the results of microarray experiments, and in this context another significance question arises. Frequently, the immediate outcome of an analysis of a microarray experiment is a long list of genes that have been doing something in the experiment; let us say that they have large interaction in a 2×2 factorial experiment. The experimenter then compares each gene on the list to the GO and finds that 15 of the genes are transcription factors, or are concerned with locomotory behavior, or are located in the nucleus. Is this number unusual, or could it have readily occurred "by chance?" A reference set could be all genes on the microarray. Biologists are interested in the answers to questions like this because they are looking for clues concerning the cellular processes affected by the experimental intervention.

It may be that questions like the two just mentioned cannot be well-posed, but it would be good to have a thorough discussion of the issues involved here.

2.6.4 Combining other data with microarray data

A class of problems of ever-increasing importance concerns the combining of microarray data with clinical data. It is easy to envisage the time when an individual's cancer diagnosis, for example, will involve a wide variety of clinical observations — the traditional diagnostic indicators — together with the absolute or relative expression levels of tens of thousands of genes. Further, we can expect such data together with clinical outcomes, such as survival data on patients. Early articles on this topic include

Alizadeh et al. (2000), Bittner et al. (2000), and Sørlie et al. (2001). It is hard to be specific here, but the challenges and opportunities appear great.

One of the common outcomes of a microarray experiment is a number of lists (perhaps clusters) of genes which are thought to be co-regulated, that is, to be acting in concert in the biological processes underlying the experiment. A very natural question for a biologist with access to the genome sequence of the organism under study is: Can we find short regulatory sequences in the genome, upstream of any these sets of genes, that might be responsible for their being co-regulated? In the jargon of molecular biology, such regulatory sequences are termed *cis*-acting, in order to distinguish them from *trans*-acting sequences, which are much more difficult to identify computationally. For some early research on this problem, see Bussemaker et al. (2000), Hughes et al. (2000a), Cohen et al. (2000), Bussemaker et al. (2001), Pilpel et al. (2001), Chiang et al. (2001), and Keles et al. (2002). A related problem is discussed in Liu et al. (2002). These articles, and others like them, are just the beginning of what will surely grow into a significant body of research that is likely to require considerable statistical input: the addressing of questions that involve both gene expression and genome sequence data. This appears to be an appropriate theme on which to close our brief summary of the design and analysis of gene expression experiments.

CHAPTER 3

Classification in microarray experiments

Sandrine Dudoit and Jane Fridlyand

Abstract. Deoxyribonucleic acid (DNA) microarrays are part of a new and promising class of biotechnologies that allow the simultaneous monitoring of expression levels in cells for thousands of genes. Microarray experiments are increasingly being performed in biological and medical research to address a wide range of problems. In cancer research, microarrays are used to study the molecular variations among tumors, with the aim of developing better diagnosis and treatment strategies for the disease. Classification is an important question in microarray experiments, for purposes of classifying biological samples and predicting clinical or other outcomes using gene expression data. Although classification is by no means a new subject in the statistical literature, the large and complex multivariate datasets generated by microarray experiments raise new methodological and computational challenges.

This chapter addresses statistical issues arising in the classification of biological samples using gene expression data from DNA microarray experiments. It discusses the statistical foundations of classification and provides an overview of different classifiers, including linear discriminant analysis, nearest neighbor classifiers, classification trees, and support vector machines. Applications of resampling methods, such as bagging and boosting, for improving classifier accuracy are described. The important questions of feature selection and classifier performance assessment are also addressed. The performance of five main types of classifiers is examined using gene expression data from recently published cancer microarray studies of breast and brain tumors.

3.1 Introduction

3.1.1 Motivation: tumor classification using gene expression data

An important problem in deoxyribonucleic acid (DNA) microarray experiments is the classification of biological samples using gene expression data. To date, this problem

1-58488-327-8/03/$0.00+$1.50
© 2003 by CRC Press LLC

has received the most attention in the context of cancer research; we thus begin this article with a review of tumor classification using microarray gene expression data. A reliable and precise classification of tumors is essential for successful diagnosis and treatment of cancer. Current methods for classifying human malignancies rely on a variety of clinical, morphological, and molecular variables. Despite recent progress, uncertainties still exist in diagnosis. Also, it is likely that the existing classes are heterogeneous and comprise diseases that are molecularly distinct and follow different clinical courses. Complementary DNA (cDNA) microarrays and high-density oligonucleotide chips are novel biotechnologies that are being used increasingly in cancer research (Alizadeh et al., 2000; Alon et al., 1999; Beer et al., 2002; Bhattacharjee et al., 2001; Bittner et al., 2000; Chen et al., 2002; Golub et al., 1999; Perou et al., 1999; Pollack et al., 1999; Pomeroy et al., 2002; Ross et al., 2000; Sørlie et al., 2001). By allowing the monitoring of expression levels in cells for thousands of genes simultaneously, microarray experiments may lead to a more complete understanding of the molecular variations among tumors, and hence to better diagnosis and treatment strategies for the disease.

Recent publications on cancer classification using gene expression data have mainly focused on the cluster analysis of both tumor samples and genes, and include applications of hierarchical clustering (Alizadeh et al., 2000; Alon et al., 1999; Bhattacharjee et al., 2001; Bittner et al., 2000; Chen et al., 2002; Perou et al., 1999; Pollack et al., 1999; Pomeroy et al., 2002; Ross et al., 2000; Sørlie et al., 2001; Tibshirani et al., 1999) and partitioning methods such as self-organizing maps (Golub et al., 1999; Pomeroy et al., 2002). Alizadeh et al. (2000) used cDNA microarray analysis of lymphoma samples to identify two previously unrecognized and molecularly distinct subclasses of diffuse large B-cell lymphomas corresponding to different stages of B-cell differentiation. One type expressed genes characteristic of germinal center B-cells (*germinal center B-like DLBCL* class), and the other expressed genes normally induced during *in vitro* activation of peripheral blood B-cells (*activated B-like DLBCL* class). They also demonstrated that patients with the two subclasses of tumors had different clinical prognoses. Average linkage hierarchical clustering was used to identify the two tumor subclasses as well as to group genes with similar expression patterns across the different samples. Similar approaches were applied by Bhattacharjee et al. (2001) to identify lung adenocarcinoma subclasses with different patient outcome. Ross et al. (2001) used cDNA microarrays to study gene expression in the 60 cell lines from the National Cancer Institute's anticancer drug screen (NCI 60). Hierarchical clustering of the cell lines based on gene expression data revealed a correspondence between gene expression and tissue of origin of the tumors. Hierarchical clustering was also used to group genes with similar expression patterns across the cell lines. Using acute leukemias as a test case, Golub et al. (1999) looked into both the cluster analysis and the discriminant analysis of tumors using gene expression data. For cluster analysis, or class discovery, self-organizing maps (SOMs) were applied to the gene expression data, and the tumor groups revealed by this method were compared to known classes. For class prediction, Golub et al. (1999) proposed a weighted gene voting scheme that turned out to be a variant of a special case of linear discriminant analysis, which is also known as naive Bayes classification. More recently,

Pomeroy et al. (2002) used Affymetrix oligonucleotide chips to study gene expression in embryonal tumors of the central nervous system (CNS). A range of unsupervised and supervised learning methods were applied to investigate whether gene expression data could be used to distinguish among new and existing CNS tumor classes and for patient prognosis. Beer et al. (2002) used an approach based on the Cox proportional-hazard model to identify marker genes for predicting the survival of patients with lung adenocarcinoma.

The recent studies just cited are instances of a growing body of research in which gene expression profiling is used to distinguish among known tumor classes, predict clinical outcomes such as survival and response to treatment, and identify previously unrecognized and clinically significant subclasses of tumors. Indeed, Golub et al. (1999) conclude that: "This experience underscores the fact that leukemia diagnosis remains imperfect and could benefit from a battery of expression-based predictors for various cancers. Most important, the technique of class prediction can be applied to distinctions relating to future clinical outcomes, such as drug response or survival." In the same vein, Alizadeh et al. (2000) "... anticipate that global surveys of gene expression in cancer, such as we present here, will identify a small number of marker genes that will be used to stratify patients into molecularly relevant categories which will improve the precision and power of clinical trials." The ability to successfully distinguish among tumor classes (already known or yet to be discovered) and to predict clinical outcomes on the basis of gene expression data is an important aspect of this novel approach to cancer classification.

Microarray experiments in cancer research are not limited to monitoring transcript or messenger ribonucleic acid (mRNA) levels. In *comparative genomic hybridization* (CGH) experiments, DNA microarrays are used to measure DNA copy numbers across the genome (Jain et al., 2001; Pollack et al., 1999; Wilhelm et al., 2002). The microarray technology can also be used to establish genome-wide DNA methylation maps. *DNA methylation* refers to the addition of a methyl group or small "tag" to the DNA molecule and can result in silencing the expression of the corresponding genes. This regulation mechanism plays an important role in diseases such as cancer (Costello et al., 2000). Similar classification questions arise in these types of microarray experiments. In addition, cancer research is only one of the many areas of application of the microarray technology. In immunology, microarrays have recently been used to study the gene expression host response to infection by bacterial pathogens (Boldrick et al., 2002). Clinical implications include improved diagnosis of bacterial infections by gene expression profiling.

The preceding examples illustrate that class prediction is an important question in microarray experiments, for purposes of classifying biological samples and predicting clinical or other outcomes using gene expression data. A closely related issue is that of feature or variable selection (i.e., the identification of marker genes that characterize different tumor classes or have good predictive power for an outcome of interest). Although classification and variable selection are by no means new subjects in the statistical literature, the large and complex multivariate datasets generated by microarray experiments raise new methodological and computational challenges.

3.1.2 Outline

This chapter discusses statistical issues arising in the classification of biological samples using gene expression data from DNA microarray experiments. The remainder of this section gives further background on classification and microarray experiments. Section 3.2 discusses the statistical foundations of classification and provides an overview of different classifiers, including linear discriminant analysis, nearest neighbor classifiers, classification trees, and support vector machines. Section 3.3 discusses general issues in classification such as feature selection, standardization, distance function, loss function, class representation, imputation of missing data, and polychotomous classification. The important question of classifier performance assessment is addressed in Section 3.4. Applications of resampling methods, such as bagging and boosting, for improving classifier accuracy are discussed in Section 3.5. Section 3.6 describes recently published gene expression datasets from cancer microarray studies of breast and brain tumors; these datasets were used to assess the performance of the classifiers introduced in Section 3.2. The results of the comparison study are reported in Section 3.7. Section 3.8 summarizes our findings and outlines open questions. Finally, Section 3.9 describes software resources for classification.

The reader is referred to the texts by Hastie et al. (2001), Mardia et al. (1979), McLachlan (1992), and Ripley (1996) for general discussions of classification. The article by Breiman (2001), with associated comments and rejoinder, addresses issues arising in modern high-dimensional classification problems. Recent work on statistical aspects of classification in the context of microarray experiments includes: Ambroise and McLachlan (2002), Chow et al. (2001), Dettling and Buhlmann (2002), Dudoit et al. (2002a), Golub et al. (1999), Moler et al. (2000), Pomeroy et al. (2002), Shieh et al. (2002), Tibshirani et al. (2002), and West et al. (2001). These articles have mostly focused on existing methods or variants thereof, and, in general, comparison studies have been limited and not always properly calibrated. Studies performed to date suggest that simple methods, such as nearest neighbor or naive Bayes classification, perform as well as more complex approaches, such as aggregated classification trees or support vector machines. The specific classifiers considered in these references are discussed in Section 3.2.

3.1.3 Background on classification

Unsupervised vs. supervised learning

In many situations, one is concerned with assigning objects to classes on the basis of measurements made on these objects. Such problems have two main aspects: discrimination and clustering, or supervised and unsupervised learning. In *unsupervised learning* (also known as *cluster analysis, class discovery*, and *unsupervised pattern recognition*), the classes are *unknown a priori* and need to be discovered from the data. This involves estimating the number of classes (or clusters) and assigning objects to these classes. In contrast, in *supervised learning* (also known as *classification, discriminant analysis, class prediction*, and *supervised pattern recognition*), the classes are *predefined* and the task is to understand the basis for the classification from a set of labeled objects (learning set). This information is used to build a classifier that will then be applied to predict the class of future unlabeled observations. In many situations, the two problems are related, because the classes that are discovered from unsupervised learning are often used at a later time in a supervised learning setting. Here, we focus on supervised learning and use the simpler term classification.

Classification

Classification is a *prediction* or *learning* problem in which the variable to be predicted assumes one of K predefined and unordered values, $\{c_1, c_2, \ldots, c_K\}$, arbitrarily relabeled by the integers $\{1, 2, \ldots, K\}$ or $\{0, 1, \ldots, K-1\}$, and sometimes $\{-1, 1\}$ in binary classification. The K values correspond to K predefined classes (e.g., tumor class, bacteria type). Associated with each object are: a *response* or *dependent variable* (class label), $Y \in \{1, 2, \ldots, K\}$, and a set of G measurements that form the *feature vector* or *vector of predictor variables*, $\mathbf{X} = (X_1, \ldots, X_G)$. The feature vector $\mathbf{X}$ belongs to a *feature space* $\mathcal{X}$ (e.g., the real numbers $\Re^G$). The task is to classify an object into one of the K classes on the basis of an observed measurement $\mathbf{X} = \mathbf{x}$ (i.e., predict Y from $\mathbf{X}$).

A *classifier* or *predictor* for K classes is a mapping $\mathcal{C}$ from $\mathcal{X}$ into $\{1, 2, \ldots, K\}$, $\mathcal{C} : \mathcal{X} \to \{1, 2, \ldots, K\}$, where $\mathcal{C}(\mathbf{x})$ denotes the predicted class for a feature vector $\mathbf{x}$. That is, a classifier $\mathcal{C}$ corresponds to a *partition* of the feature space $\mathcal{X}$ into K disjoint and exhaustive subsets, $A_1, \ldots, A_K$, such that a sample with feature vector $\mathbf{x} = (x_1, \ldots, x_G) \in A_k$ has predicted class $\hat{y} = k$. Modifications can be made to allow doubt or outlier classes (Ripley, 1996).

Classifiers are built or *trained* from past experience (i.e., from observations that are known to belong to certain classes). Such observations comprise the *learning set* (LS), $\mathcal{L} = \{(\mathbf{x}_1, y_1), \ldots, (\mathbf{x}_n, y_n)\}$. A classifier built from a learning set $\mathcal{L}$ is denoted by $\mathcal{C}(\cdot; \mathcal{L})$. When the learning set is viewed as a collection of random variables, the resulting classifier is also a *random variable*. Intuitively, for a fixed value of the feature vector $\mathbf{x}$, as the learning set varies, so will the predicted class $\mathcal{C}(\mathbf{x}; \mathcal{L})$. It is thus meaningful to consider distributional properties (e.g., bias and variance) of classifiers when assessing or comparing the performance of different classifiers. Such properties are discussed in Section 3.4.1.

Classification for gene expression data

In the case of gene expression data from cancer DNA microarray experiments, features correspond to the expression measures of different genes and classes correspond to different tumor types (e.g., nodal positive vs. negative breast tumors, or tumors with good vs. bad prognosis). Three main types of statistical problems are associated with tumor classification: (i) identification of new tumor classes using gene expression profiles — *unsupervised learning*; (ii) classification of malignancies into known classes — *supervised learning*; (iii) identification of marker genes that characterize the different tumor classes — *feature selection.* Unsupervised learning is the subject of Chapter 4 of this volume; statistical inference issues related to the cluster analysis of genes and samples are discussed in van der Laan and Bryan (2001) and Pollard and van der Laan (2002). This chapter focuses primarily on the second statistical problem discussed previously and briefly addresses the related issue of feature selection.

For our purposes, gene expression data on G genes for n tumor mRNA samples may be summarized by a $G \times n$ matrix $X = (x_{gi})$, where x_{gi} denotes the expression measure of gene (variable) g in mRNA sample (observation) i. (Note that this gene expression data matrix is the transpose of the standard $n \times G$ design matrix. The $G \times n$ representation was adopted in the microarray literature for display purposes, because for very large G and small n it is easier to display a $G \times n$ matrix than an $n \times G$ matrix.) The expression levels might be either absolute (e.g., oligonucleotide arrays used to produce the breast and brain tumor datasets of Section 3.6) or relative to the expression levels of a suitably defined common reference sample (e.g., cDNA microarrays used in Alizadeh et al., 2000). When the mRNA samples belong to known classes (e.g., ER positive or negative samples for breast tumors), the data for each observation consist of a *gene expression profile* $\mathbf{x}_i = (x_{1i}, \ldots, x_{Gi})$ and a class label y_i (i.e., of predictor variables $\mathbf{x}_i$ and response y_i). For K tumor classes, the class labels y_i are defined to be integers ranging from 1 to K, and n_k denotes the number of learning set observations belonging to class k. Note that the expression measures x_{gi} are, in general, highly processed data: the raw data in a microarray experiment consist of image files, and important preprocessing steps include image analysis of the scanned images and normalization (Dudoit and Yang, 2003; Irizarry et al., 2002; Yang et al., 2001, 2002). Preprocessing is also discussed in Chapter 1 of this book. Data from these new types of experiments present a so-called "small n, large p" problem, that is, a very large number of variables (genes) relative to the number of observations (tumor samples). The publicly available datasets typically contain expression data on 5000–10,000 genes for less than 100 tumor samples. Both numbers are expected to grow, with the number of genes reaching on the order of 30,000, which is an estimate for the total number of genes in the human genome.

Many different approaches are available to build a classifier for tumor samples using gene expression data; various methods are reviewed in Section 3.2. Different classifiers will clearly have varying performance (i.e., different error rates). In addition, decisions concerning, for example, the set of genes used to build the classifier (i.e., feature selection) and the standardization of the expression data can have a large impact on classifier performance (see Section 3.3). In the context of tumor classification, errors could correspond to misdiagnosis and possibly assignment to improper treatment protocol. Thus, an essential task is to assess the accuracy of the classifier. Distributional properties of classifiers and performance assessment are discussed in Section 3.4.

3.2 Overview of different classifiers

3.2.1 Classification and statistical decision theory

It is useful to view classification as a *statistical decision theory* problem. For each object, a feature vector $\mathbf{x}$ is examined to decide to which of a fixed set of classes that object belongs. Assume observations are independently and identically distributed (i.i.d.) from an unknown multivariate distribution. Denote the class k *prior*, i.e., the proportion of objects of class k in the population of interest, by $\pi_k = p(Y = k)$. Objects in class k have feature vectors with *class conditional density* $p_k(\mathbf{x}) = p(\mathbf{x}|Y = k)$. Define a *loss function* L, where $L(h,l)$ elaborates the loss incurred if a class h case is erroneously classified as belonging to class l (see Section 3.3.3). The *risk function* for a classifier $\mathcal{C}$ is the expected loss when $\mathcal{C}$ is used to classify, that is,

$$R(\mathcal{C}) = E[L(Y,\mathcal{C}(\mathbf{X}))] = \sum_k E[L(k,\mathcal{C}(\mathbf{X}))|Y = k]\pi_k = \sum_k \int L(k,\mathcal{C}(\mathbf{x}))p_k(\mathbf{x})\pi_k d\mathbf{x}. \quad (3.1)$$

Typically, $L(h,h) = 0$, and in many cases the loss is defined to be symmetric with $L(h,l) = 1$, $h \neq l$, i.e., making an error of one type is equivalent to making an error of a different type. Then, the risk is simply the *misclassification rate*, $p(\mathcal{C}(\mathbf{X}) \neq Y) = \sum_k \int_{\mathcal{C}(\mathbf{x})\neq k} p_k(\mathbf{x})\pi_k d\mathbf{x}$. Note that, here, the classifier is viewed as fixed; that is, if a learning set $\mathcal{L}$ is used to train the classifier, probabilities are conditional on $\mathcal{L}$. When (unrealistically) both π_k and $p_k(\mathbf{x})$ are known, it is possible to define an optimal classifier which minimizes the risk function. This situation gives an upper bound on the performance of classifiers in the more realistic setting where these distributions are unknown (see Bayes rule and Bayes risk discussed next).

The Bayes rule

In the unlikely situation that the class conditional densities $p_k(\mathbf{x}) = p(\mathbf{x}|Y = k)$ and class priors π_k are known, Bayes' theorem may be used to express the *posterior probability* $p(k \mid \mathbf{x})$ of class k given feature vector $\mathbf{x}$ as

$$p(k \mid \mathbf{x}) = \frac{\pi_k p_k(\mathbf{x})}{\sum_l \pi_l p_l(\mathbf{x})}.$$

The *Bayes rule* predicts the class of an observation $\mathbf{x}$ by that with highest posterior probability

$$\mathcal{C}_B(\mathbf{x}) = \text{argmax}_k\, p(k \mid \mathbf{x}).$$

The class posterior probabilities reflect the confidence in predictions for individual observations; the closer they are to one, the greater the confidence. The Bayes rule minimizes the risk function or misclassification rate under a symmetric loss function — *Bayes risk.* For a general loss function L, the classification rule which minimizes the risk function is

$$\mathcal{C}_B(\mathbf{x}) = \text{argmin}_l \sum_{h=1}^{K} L(h,l)p(h \mid \mathbf{x}). \tag{3.2}$$

In the special case when $L(h,l) = L_h I(h \neq l)$, that is, the loss incurred from misclassifying a class h observation is the same irrespective of the predicted class l, the Bayes rule is

$$\mathcal{C}_B(\mathbf{x}) = \text{argmax}_k \, L_k \, p(k \mid \mathbf{x}). \tag{3.3}$$

Suitable adjustments can be made to accommodate the doubt and outlier classes (see Ripley, 1996, p. 19).

Many classifiers can be viewed as versions of this general rule, with particular parametric or nonparametric estimators of $p(k \mid \mathbf{x})$. Two general paradigms are used to estimate the class posterior probabilities $p(k \mid \mathbf{x})$: the density estimation and the direct function estimation paradigms (Friedman, 1996b). In the *density estimation* approach, class conditional densities $p_k(\mathbf{x}) = p(\mathbf{x}|Y = k)$ (and priors π_k) are estimated separately for each class and Bayes' theorem is applied to obtain estimates of $p(k \mid \mathbf{x})$. Classification procedures employing density estimation include: Gaussian maximum likelihood discriminant rules, also known as discriminant analysis (Chapter 3 in Ripley, 1996 and Section 3.2.3 in this book); learning vector quantization (Section 6.3 in Ripley, 1996), and Bayesian belief networks (Chapter 8 in Ripley, 1996). Another example is given by *naive Bayes methods*, which approximate class conditional densities $p_k(\mathbf{x})$ by the product of their marginal densities on each feature. In the *direct function estimation* approach, class posterior probabilities $p(k \mid \mathbf{x})$ are estimated directly based on function estimation methodology such as regression. This paradigm is used by popular classification procedures such as: logistic regression (Chapter 3 in Ripley, 1996 and Section 3.2.4 in this book), neural networks (Chapter 5 in Ripley, 1996), classification trees (Breiman et al., 1984 and Section 3.2.6 in this book), projection pursuit (Section 6.1 in Ripley, 1996), and nearest neighbor classifiers (Section 6.2 in Ripley, 1996 and Section 3.2.5 in this book).

Maximum likelihood discriminant rules

The frequentist analogue of the Bayes rule is the maximum likelihood discriminant rule. For known class conditional densities $p_k(\mathbf{x}) = p(\mathbf{x}|Y = k)$, the *maximum likelihood* (ML) discriminant rule predicts the class of an observation $\mathbf{x}$ by that which gives the largest likelihood to $\mathbf{x}$: $\mathcal{C}(\mathbf{x}) = \text{argmax}_k \, p_k(\mathbf{x})$. In the case of equal class priors π_k, this amounts to maximizing the class posterior probabilities $p(k|\mathbf{x})$, i.e., the Bayes rule. Otherwise, the ML rule is not optimal, in the sense that it does not minimize the risk function.

3.2.2 Fisher linear discriminant analysis

First applied by Barnard (1935) at the suggestion of Fisher (1936), *Fisher linear discriminant analysis* (FLDA) is based on finding linear combinations $\mathbf{xa}$ of the $1 \times G$ feature vectors $\mathbf{x} = (x_1, \ldots, x_G)$ with large ratios of between-groups to within-groups sums of squares (Mardia et al., 1979) for a detailed presentation of FLDA). For a $G \times n$ learning set data matrix X, the linear combination $\mathbf{a}'X$ of the rows of X has ratio of between-groups to within-groups sums of squares given by $\mathbf{a}'B\mathbf{a}/\mathbf{a}'W\mathbf{a}$, where B and W denote respectively the $G \times G$ matrices of between-groups and within-groups sums of squares and cross-products. The extreme values of $\mathbf{a}'B\mathbf{a}/\mathbf{a}'W\mathbf{a}$ are obtained from the eigenvalues and eigenvectors of $W^{-1}B$. The matrix $W^{-1}B$ has at most $s = \min(K-1, G)$ nonzero eigenvalues, $\lambda_1 \geq \lambda_2 \geq \ldots \geq \lambda_s$, with corresponding linearly independent eigenvectors $\mathbf{v}_1, \mathbf{v}_2, \ldots, \mathbf{v}_s$. The *discriminant variables* are defined to be $\mathbf{xv}_l$, $l = 1, \ldots, s$, and, in particular, $\mathbf{a} = \mathbf{v}_1$ maximizes $\mathbf{a}'B\mathbf{a}/\mathbf{a}'W\mathbf{a}$.

For a feature vector $\mathbf{x} = (x_1, \ldots, x_G)$, let $d_k^2(\mathbf{x}) = \sum_{l=1}^{s}((\mathbf{x} - \bar{\mathbf{x}}_k)\mathbf{v}_l)^2$ denote its (squared) Euclidean distance, in terms of the discriminant variables, from the $1 \times G$ vector of class k sample means $\bar{\mathbf{x}}_k = (\bar{x}_{k1}, \ldots, \bar{x}_{kG})$ for the learning set $\mathcal{L}$ (here, $\bar{x}_{kg} = \sum_i I(y_i = k)x_{gi}/n_k$). The predicted class for feature vector $\mathbf{x}$ is the class whose mean vector $\bar{\mathbf{x}}_k$ is closest to $\mathbf{x}$ in the space of discriminant variables, i.e., $\mathcal{C}(\mathbf{x}; \mathcal{L}) = \text{argmin}_k\ d_k(\mathbf{x})$. Thus, the two main steps in FLDA are: a feature selection or dimensionality reduction step leading to the identification of s discriminant variables; and a classification step per se, in which observations are classified in terms of their distances from class means in the reduced space. FLDA is a non-parametric method which also arises in a parametric setting. For $K = 2$ classes, FLDA yields the same classifier as the sample maximum likelihood discriminant rule for multivariate Gaussian class densities with the same covariance matrix (see Section 3.2.3, LDA for $K = 2$).

3.2.3 Linear and quadratic discriminant analysis

Linear and quadratic (in the features $\mathbf{x}$) discriminant rules arise as Bayes rules or maximum likelihood (ML) discriminant rules when features have Gaussian distributions within each class. Specifically, let $\mathbf{X}|Y = k \sim N(\mu_k, \Sigma_k)$, where $\mu_k = (\mu_{k1}, \ldots, \mu_{kG})$ and Σ_k denote respectively the expected value and the $G \times G$ covariance matrix of the feature vector in class k. The Bayes rule is

$$\mathcal{C}(\mathbf{x}) = \text{argmin}_k \left\{ (\mathbf{x} - \mu_k)\Sigma_k^{-1}(\mathbf{x} - \mu_k)' + \log|\Sigma_k| - 2\log \pi_k \right\}. \qquad (3.4)$$

In general, this is a quadratic discriminant rule — *quadratic discriminant analysis* (QDA). The main quantity in the discriminant rule is $(\mathbf{x} - \mu_k)\Sigma_k^{-1}(\mathbf{x} - \mu_k)'$, the squared Mahalanobis distance from the observation $\mathbf{x}$ to the class k mean vector μ_k. Interesting special cases are described next for homogeneous priors (i.e., for π_k constant in k).

Linear discriminant analysis (LDA)

When the class densities have the same covariance matrix, $\Sigma_k = \Sigma$, the discriminant rule is based on the square of the Mahalanobis distance and is linear in $\mathbf{x}$ and given by $\mathcal{C}(\mathbf{x}) = \text{argmin}_k \ (\mathbf{x} - \mu_k)\Sigma^{-1}(\mathbf{x} - \mu_k)' = \text{argmin}_k \ \left(\mu_k\Sigma^{-1}\mu_k' - 2\mathbf{x}\Sigma^{-1}\mu_k'\right)$.

Diagonal quadratic discriminant analysis (DQDA)

When the class densities have diagonal covariance matrices, $\Delta_k = \text{diag}(\sigma_{k1}^2, \ldots, \sigma_{kG}^2)$, the discriminant rule is given by additive quadratic contributions from each feature, that is, $\mathcal{C}(\mathbf{x}) = \text{argmin}_k \sum_{g=1}^{G} \left\{ \frac{(x_g - \mu_{kg})^2}{\sigma_{kg}^2} + \log \sigma_{kg}^2 \right\}$.

Diagonal linear discriminant analysis (DLDA)

When the class densities have the same diagonal covariance matrix $\Delta = \text{diag}(\sigma_1^2, \ldots, \sigma_G^2)$, the discriminant rule is linear and given by $\mathcal{C}(\mathbf{x}) = \text{argmin}_k \sum_{g=1}^{G} \frac{(x_g - \mu_{kg})^2}{\sigma_g^2}$.

Nearest centroid

In this simplest case, it is assumed that $\Sigma_k = I_G$, the $G \times G$ identity matrix. Observations are then classified on the basis of their Euclidean distance from class means.

DLDA and DQDA correspond to *naive Bayes rules* for Gaussian class conditional densities. As with any classifier explicitly estimating the Bayes rule, class posterior probabilities may be used to assess the confidence in predictions for individual observations. Note that, although LDA and QDA were introduced within a parametric context as ML or Bayes rules for Gaussian class densities, these methods have much wider applicability.

For the sample Bayes or ML discriminant rules, the population mean vectors and covariance matrices are estimated from a learning set $\mathcal{L}$, by the sample mean vectors and covariance matrices, respectively: $\hat{\mu}_k = \bar{\mathbf{x}}_k$ and $\hat{\Sigma}_k = S_k$. For the constant covariance matrix case, the pooled estimator of the common covariance matrix is used: $\hat{\Sigma} = S = \sum_k (n_k - 1)S_k/(n - K)$.

The preceding simple rules may be modified easily to allow unequal class priors; estimates of the priors may be obtained from the sample class proportions $\hat{\pi}_k = n_k/n$. A compromise between LDA and QDA is *regularized discriminant analysis*, which shrinks the covariance matrices S_k of QDA toward a common covariance matrix S as in LDA: $S_k(\alpha) = \alpha S_k + (1 - \alpha)S$, where the parameter α can be chosen to optimize prediction accuracy, using, for example, cross-validation (see Section 3.4 in Ripley, 1996 and Section 4.3.1 in Hastie et al., 2001).

Golub et al. weighted gene voting scheme

In one of the first applications of a classification method to gene expression data, Golub et al. (1999) proposed a *weighted gene voting scheme* for binary classification. This method turns out to be a variant of DLDA or naive Bayes classification. For two classes, $k = 1$ and 2, the sample DLDA rule assigns a case with gene expression profile $\mathbf{x} = (x_1, \ldots, x_G)$ to class 1 if and only if

$$\sum_{g=1}^{G} \frac{(x_g - \bar{x}_{2g})^2}{\hat{\sigma}_g^2} \geq \sum_{g=1}^{G} \frac{(x_g - \bar{x}_{1g})^2}{\hat{\sigma}_g^2},$$

that is,

$$\sum_{g=1}^{G} \frac{(\bar{x}_{1g} - \bar{x}_{2g})}{\hat{\sigma}_g^2} \left(x_g - \frac{(\bar{x}_{1g} + \bar{x}_{2g})}{2} \right) \geq 0.$$

The discriminant function can be rewritten as $\sum_g v_g$, where $v_g = a_g(x_g - b_g)$, $a_g = (\bar{x}_{1g} - \bar{x}_{2g})/\hat{\sigma}_g^2$, and $b_g = (\bar{x}_{1g} + \bar{x}_{2g})/2$. This is almost the same function as used in Golub et al., except for a_g that Golub et al. (1999) define as $a_g = (\bar{x}_{1g} - \bar{x}_{2g})/(\hat{\sigma}_{1g} + \hat{\sigma}_{2g})$. The quantity $\hat{\sigma}_{1g} + \hat{\sigma}_{2g}$ is an unusual estimator of the standard error of a difference and having standard deviations instead of variances in the denominator of a_g produces the wrong units for the discriminant function. For each prediction made by the classifier, Golub et al. (1999) also define a prediction strength, PS, which indicates the "margin of victory": $PS = (\max(V_1, V_2) - \min(V_1, V_2))/(\max(V_1, V_2) + \min(V_1, V_2))$, where $V_1 = \sum_g \max(v_g, 0)$ and $V_2 = \sum_g \max(-v_g, 0)$. Golub et al. (1999) choose a conservative prediction strength threshold of .3 below which no predictions are made. Note that as implemented in Golub et al. (1999), the weighted gene voting scheme only allows binary classification. However, if viewed as a variant of DLDA described previously it can be extended easily to polychotomous classification problems and to accommodate different class prior probabilities.

3.2.4 Extensions of linear discriminant analysis

The linear discriminant rules described previously are classical and widely used classification tools. Features explaining their popularity include: (i) simple and intuitive rule — the predicted class of a test case is the class with the closest mean (using the Mahalanobis metric); (ii) estimated Bayes rule — LDA is based on the estimated Bayes rule for Gaussian class conditional densities and homogeneous class priors; (iii) easy to implement — the partition corresponding to the classifier has linear boundaries; (iv) good performance in practice — despite a possibly high bias, the low variance of the naive Bayes estimators of class posterior probabilities often results in low classification error (Friedman, 1996b). However, LDA has a number of obvious limitations that are due in part to its simplicity: (i) linear or even quadratic discriminant boundaries may not be flexible enough; (ii) features may have mixture distributions within classes (related to (i)); (iii) for a large number of features, performance may degrade rapidly due to over-parameterization and high variance parameter estimators. Hastie et al. (2001) describe three main classes of extensions to LDA to address these shortcomings. A modification of DLDA was recently considered by Tibshirani et al. (2002) in the context of microarray data analysis.

Flexible discriminant analysis (FDA)

Viewing LDA as a linear regression problem, suggests considering more general non-parametric forms of regression for building a more flexible classifier. FDA amounts to performing LDA on transformed responses or features (Section 12.5 in Hastie et al., 2001). *Logistic discrimination* is one such extension that can be motivated as follows. For Gaussian class conditional densities with common covariance matrix (and other models for the feature distributions), posterior probabilities $p(k|\mathbf{x})$ satisfy $\log p(k|\mathbf{x}) - \log p(1|\mathbf{x}) = \alpha_k + \mathbf{x}\beta_k$. This suggests modeling $\log p(k|\mathbf{x}) - \log p(1|\mathbf{x})$ more generally by some parametric family of functions, say $g_k(\mathbf{x};\theta)$ with $g_1(\mathbf{x};\theta) \equiv 0$. Estimates of the class posterior probabilities are then given by

$$\hat{p}(k|\mathbf{x}) = \frac{\exp g_k(\mathbf{x};\hat{\theta})}{\sum_l \exp g_l(\mathbf{x};\hat{\theta})}.$$

Classification is done by the (estimated) Bayes rule, i.e., $\mathcal{C}(\mathbf{x};\mathcal{L}) = \text{argmax}_k\, \hat{p}(k|\mathbf{x})$. In the machine learning literature, the function $\exp a_k / \sum_l \exp a_l$ is known as the *softmax function*. In the earlier statistical literature, it is known as the *multiple logit function*. In the linear case, one takes $g_k(\mathbf{x};\theta) = \alpha_k + \mathbf{x}\beta_k$. Logistic regression models are typically fit by maximum likelihood, using a Newton–Raphson algorithm known as *iteratively reweighted least squares* (IRLS). Logistic discrimination provides a more direct way of estimating posterior probabilities and is also easier to generalize than classical linear discriminant analysis (LDA and QDA), as seen, for example, with neural networks (Chapter 8 in Ripley, 1996).

Penalized discriminant analysis (PDA)

A penalized Mahalanobis distance is used to enforce smoothness of the within-class covariance matrix of the features, possibly transformed as in FDA (Section 12.6 in Hastie et al., 2001).

Mixture discriminant analysis (MDA)

Class conditional densities are modeled as mixtures of Gaussian densities, with different mean vectors but the same covariance matrices. The EM algorithm may be used for maximum likelihood estimation of the parameters of the Gaussian components (Section 12.7 in Hastie et al., 2001).

Shrinkage methods

In a recent article, Tibshirani et al. (2002) propose an extension of the nearest centroid classifier, which allows automatic feature selection for each class separately. This approach is used to deal with the very large number of features in classification problems involving microarray data. *Nearest shrunken centroids* is a modification of nearest centroids described earlier with: shrunken class mean vectors, based on the value of a t-like test statistic; addition of a positive constant s_0 to the variance $\hat{\sigma}_g^2$ of each gene (cf. Equation (6) in Tibshirani et al., 2002). The parameter Δ controlling the amount of shrinking is selected by tenfold cross-validation; the case $\Delta = 0$ corresponds to the nearest centroid classifier. Shrinking means (and covariance matrices) in the context of discriminant analysis is also discussed in Section 3.4 of Ripley (1996). Use of shrunken mean vectors allows for automatic selection of a subset of genes that discriminate between the classes.

3.2.5 Nearest neighbor classifiers

Nearest neighbor methods are based on a *distance function* for pairs of observations, such as the Euclidean distance or one minus the Pearson correlation of their feature vectors (see Section 3.3.2 and Table 3.1 for a description of distance functions). The basic *k-nearest neighbor* (k-NN) rule proceeds as follows to classify a new observation on the basis of the learning set: find the k closest observations in the learning set and predict the class by majority vote (i.e., choose the class that is most common among those k neighbors). k-nearest neighbor classifiers suggest simple estimates of the class posterior probabilities. More formally, let $d_i(\mathbf{x}) = d(\mathbf{x}, \mathbf{x}_i)$ denote the distance between learning set observation $\mathbf{x}_i$ and the test case $\mathbf{x}$, and denote the sorted distances (order statistics) by $d_{(1)}(\mathbf{x}) \leq \ldots \leq d_{(n)}(\mathbf{x})$. Then, class posterior probability estimates $\hat{p}(l|\mathbf{x})$ for the standard k-nearest neighbor classifier are given by the fraction of class l observations among the k closest neighbors to the test case

$$\hat{p}(l|\mathbf{x}) = \frac{1}{k} \sum_{i=1}^{n} I(d_i(\mathbf{x}) \leq d_{(k)}(\mathbf{x}))\, I(y_i = l).$$

The decision rule is then $\mathcal{C}(\mathbf{x}; \mathcal{L}) = \text{argmax}_l \ \hat{p}(l|\mathbf{x})$; this can be viewed as direct estimation of the Bayes rule. Note that when a large enough number of neighbors k is used, the class posterior probability estimates $\hat{p}(l|\mathbf{x})$ may be used to measure confidence for individual predictions.

Nearest neighbor classifiers were initially proposed by Fix and Hodges (1951) as consistent nonparametric estimators of maximum likelihood discriminant rules. Nonparametric estimators of the class conditional densities $p_k(\mathbf{x})$ are obtained by first reducing the dimension of the feature space $\mathcal{X}$ from G to one using a distance function. The proportions of neighbors in each class are then used in place of the corresponding class conditional densities in the maximum likelihood discriminant rule.

Table 3.1 *Distance functions*

Name	Formula
Euclidean metric	$d_E(\mathbf{x}_i, \mathbf{x}_j) = \{\sum_g w_g (x_{gi} - x_{gj})^2\}^{1/2}$
Unstandardized	$w_g = 1$
Standardized by s.d. (Karl Pearson metric)	$w_g = 1/s_g^2$
Standardized by range	$w_g = 1/R_g^2$
Mahalanobis metric	$d_{Ml}(\mathbf{x}_i, \mathbf{x}_j) = \{(\mathbf{x_i} - \mathbf{x_j}) S^{-1} (\mathbf{x_i} - \mathbf{x_j})'\}^{1/2}$ $= \{\sum_g \sum_{g'} s_{gg'}^{-1} (x_{gi} - x_{gj})(x_{g'i} - x_{g'j})\}^{1/2}$ where $S = (s_{gg'})$ is any $G \times G$ positive definite matrix, usually the sample covariance matrix of the variables; when $S = I_G$, d_{Ml} reduces to the unstandardized Euclidean metric.
Manhattan metric	$d_{Mn}(\mathbf{x}_i, \mathbf{x}_j) = \sum_g w_g \lvert x_{gi} - x_{gj} \rvert$
Minkowski metric	$d_{Mk}(\mathbf{x}_i, \mathbf{x}_j) = \{\sum_g w_g \lvert x_{gi} - x_{gj} \rvert^\lambda\}^{1/\lambda}$, $\lambda \geq 1$ $\lambda = 1$: Manhattan metric $\lambda = 2$: Euclidean metric
Canberra metric	$d_C(\mathbf{x}_i, \mathbf{x}_j) = \sum_g \frac{\lvert x_{gi} - x_{gj} \rvert}{\lvert x_{gi} + x_{gj} \rvert}$
One-minus-Pearson-correlation	$d_{corr}(\mathbf{x}_i, \mathbf{x}_j)$ $= 1 - \frac{\sum_g (x_{gi} - \bar{x}_{.i})(x_{gj} - \bar{x}_{.j})}{\{\sum_g (x_{gi} - \bar{x}_{.i})^2\}^{1/2} \{\sum_g (x_{gj} - \bar{x}_{.j})^2\}^{1/2}}$

Note: A *metric* d satisfies the following five properties: (i) nonnegativity $d(a, b) \geq 0$; (ii) symmetry $d(a, b) = d(b, a)$; (iii) $d(a, a) = 0$; (iv) definiteness $d(a, b) = 0$ if and only if $a = b$; (v) triangle inequality $d(a, b) + d(b, c) \geq d(a, c)$. Here, the term *distance* refers to a function that is required to satisfy the first three properties only. The formulae refer to distances between observations (arrays).

Immediate questions concerning k-NN classifiers are: How should we choose the distance function d? Which features should we use? How should we choose the number of neighbors k? How should we handle tied votes? Should we weight votes based on distance? Should we require a minimum of consensus (e.g., at least l out of k votes for the winning class)? Some of these issues are discussed next, while others are discussed in a more general context in Section 3.3. The reader is referred to Section 6.2 in Ripley (1996) and to Section 13.3 through Section 13.5 in Hastie et al. (2001) for more detailed discussions of nearest neighbor methods.

Number of neighbors k

The $k = 1$ nearest neighbor partition of the feature space $\mathcal{X}$ corresponds to the *Dirichlet tessellation* of the learning set. Although classifiers with $k = 1$ are often quite successful, the number of neighbors k can have a large impact on the performance of the classifier and should be chosen carefully. A common approach for selecting the number of neighbors is leave-one-out cross-validation. Each observation in the learning set is treated in turn as if its class were unknown: its distance to all of the other learning set observations (except itself) is computed, and it is classified by the nearest neighbor rule. The classification for each learning set observation is then compared to the truth to produce the cross-validation error rate. This is done for a number of k's (e.g., $k \in \{1, 3, 5, 7\}$) and the k for which the cross-validation error rate is smallest is retained.

The nearest neighbor rule can be refined and extended to deal with unequal class priors, differential misclassification costs, and feature selection. Many of these modifications involve some form of *weighted voting* for the neighbors, where weights reflect priors and costs.

Class priors

If the class prior probabilities π_l are known and the class sampling probabilities for the learning set are different from the π_l, votes may need to be weighted according to neighbor class. That is, if n_l denotes the learning set frequency of class l, then the weights for class l neighbors may be chosen as $w_l = \pi_l / n_l$. Another approach is to weight the distance as in Brown and Koplowitz (1979) by $(n_l/\pi_l)^G$ for a G-dimensional feature space. See also Section 3.3.4.

Loss function

If differential misclassification costs occur, the minimum vote majority may be adjusted based on the class to be called. This corresponds to weighted voting by the neighbors, where weights are suggested by Equation (3.2). For example, when $L(h, l) = L_h I(h \neq l)$, that is, the loss for misclassifying a class h observation is the same irrespective of the predicted class, the weights are given by $w_l = L_l$.

Distance weights

The standard k-nearest neighbor rule equally weights all k neighbors, regardless of their distance from the test case. A more sensitive rule may be obtained by assigning weights to the neighbors that are inversely proportional to their distance from the test case. However, distance-weighting the votes has proved controversial (see Ripley, 1996, p. 198). Pomeroy et al. (2002) applied weighted k-NN to brain tumor microarray data.

Feature selection

One of the most important issues in k-nearest neighbor classification is the choice of a distance function and the selection of relevant features. For instance, in the context of microarray experiments, a large number of genes are constantly expressed across classes, thus resulting in a large number of irrelevant or noise variables. Inclusion of features with little or no relevance can substantially degrade the performance of the classifier. Feature selection may be implemented as part of the classification rule itself, by, for example, modifying the distance function used in k-NN or using classification trees. Friedman (1994) proposed a flexible metric nearest neighbor classification approach, the *machette*, in which the relevance of each feature (or linear combination of features) is estimated locally for each test case. The machette is a hybrid between classical nearest neighbor classifiers and tree-structured recursive partitioning techniques. Motivated by Friedman's (1994) work, Hastie and Tibshirani (1996b) suggested a *discriminant adaptive nearest neighbor* (DANN) procedure, in which the distance function is based on local discriminant information. More recently, Buttrey and Karo (2002) proposed a hybrid or composite classifier, *k-NN-in-leaf*, which partitions the feature space using a classification tree and classifies test set cases using a standard k-nearest neighbor rule applied only to cases in the same leaf as the test case. Each k-NN classifier may have a different number of neighbors k, a different set of features, and a different choice of scaling. The k-NN-in-leaf classifier differs from Friedman's machette in that only one tree is ever grown.

k-NN classifiers are often criticized because of their heavy computing time and storage requirements for large learning sets. Various *data editing* approaches, which retain only a small portion of the learning set, have been suggested to improve computing efficiency (see Ripley, 1996, p. 198).

3.2.6 Classification trees

Binary tree structured classifiers are constructed by repeated splits of subsets, or *nodes*, of the feature space $\mathcal{X}$ into two descendant subsets, starting with $\mathcal{X}$ itself. Each terminal subset is assigned a class label and the resulting partition of $\mathcal{X}$ corresponds to the classifier. Tree construction consists of three main aspects:

1. Selection of the splits,
2. Decision to declare a node terminal or to continue splitting,
3. Assignment of each terminal node to a class.

Different tree classifiers use different approaches to deal with these three issues. Here, we use CART — *classification and regression trees* — of Breiman et al. (1984). Other tree classifiers are C4.5, QUEST, and FACT; an extensive comparison study is found in Lim et al. (2000).

Splitting rule

Details about the splitting rule are presented in Chapter 2 and Chapter 4 of Breiman et al. (1984). The simplest splits are based on the value of a single variable. The main idea is to split a node so that the data in each of the descendant subsets are *purer* than the data in the parent subset. A number of definitions are needed in order to provide a precise definition of a node splitting rule. An *impurity function* is a function $\phi(\cdot)$ defined on the set of all K-tuples $\mathbf{p} = (p_1, \ldots, p_K)$, with $p_k \geq 0$, $k = 1, \ldots, K$, and $\sum_k p_k = 1$. It has the following properties:

1. $\phi(\mathbf{p})$ is maximal if $\mathbf{p}$ is uniform (i.e., $p_k = 1/K$ for all k).
2. $\phi(\mathbf{p})$ is zero if $\mathbf{p}$ is concentrated on one class (i.e., $p_k = 1$ for some k).
3. ϕ is symmetric in $\mathbf{p}$ (i.e., invariant to permutations of the entries p_k).

Commonly used measures of impurity include the *Gini index*, $\phi(\mathbf{p}) = \sum_{k \neq l} p_k p_l = 1 - \sum_k p_k^2$, and the *entropy*, $\phi(\mathbf{p}) = -\sum_k p_k \log p_k$ (where $0 \log 0 \equiv 0$). CART uses the Gini criterion as a default to grow classification trees.

For a node t, let $n(t)$ denote the total number of learning set cases in t and $n_k(t)$ the number of class k cases in t. For class priors π_k, the resubstitution estimate of the probability that a case belongs to class k and falls into node t is given by $\hat{p}(k, t) = \pi_k n_k(t)/n_k$. The resubstitution estimate $\hat{p}(t)$ of the probability that a case falls into node t is defined by $\hat{p}(t) = \sum_k \hat{p}(k, t)$, and the resubstitution estimate of the conditional probability that a case at node t belongs to class k is $\hat{p}(k|t) = \hat{p}(k, t)/\hat{p}(t)$. When data priors $\pi_k = n_k/n$ are used, $\hat{p}(k|t)$ is simply the relative proportion of class k cases in node t, $n_k(t)/n(t)$. Define the *impurity measure* $i(t)$ of node t by

$$i(t) = \phi(\hat{p}(1|t), \ldots, \hat{p}(K|t)).$$

Having defined node impurities, we are now in a position to define a *splitting rule*. Suppose a split s of a parent node t sends a proportion p_R of the cases in t to the right daughter node t_R and p_L to the left daughter node t_L. Then, the goodness of split is measured by the *decrease in impurity*

$$\Delta i(s, t) = i(t) - p_R i(t_R) - p_L i(t_L).$$

The split s that provides the largest improvement $\Delta i(s, t)$ is used to split node t and is called the *primary split*. Splits that are nearly as good as the primary split are called *competitor splits*. Finally, *surrogate splits* are defined as splits that most closely imitate the primary split. Informally, "imitation" means that if a surrogate split is used instead of a primary split, the resulting daughter nodes will be very similar to the ones defined by the primary split. The split that provides the best agreement of the two sets of daughter nodes is called the *first surrogate split*. Surrogate splits are very useful for handling missing data (see Section 3.3.5).

Split-stopping rule

Details about the split-stopping rule are presented in Chapter 3 of Breiman et al. (1984). Briefly, the right-sized tree and accurate estimates of classification error can be obtained as follows. A large tree is grown and selectively *pruned* upward, yielding a decreasing sequence of subtrees. Cross-validation is then used to identify the subtree having the lowest estimated misclassification rate.

Class assignment rule

Details about the class assignment rule are presented in Chapter 2 of Breiman et al. (1984). For each terminal node, choose the class that minimizes the resubstitution estimate of the misclassification probability, given that a case falls into this node. Note that given equal costs and priors, the resulting class is simply the majority class in that node.

A number of refinements to the basic CART procedure are described next. These include specification of two related quantities, the class priors π_k and loss function $L(h,l)$. (Note that the CART notation in Chapter 4 of Breiman et al. (1984) is $L(h,l) = C(l|h)$ for the cost of misclassifying a class h observation as a class l observation.) One can alter class priors and the loss function using the CART version 1.310 `option` file. For the R `rpart()` function, this would be done by modifying the `prior` and `loss` components of the argument `parms`.

Class priors

In some studies, the learning set may be very unbalanced between classes, or a rare class may be proportionally over-represented to reduce estimation biases without having a large learning set size n (see Section 3.3.4). Thus, it is often necessary to use priors other than the sampling proportions of individual classes. In general, putting a large prior on a class tends to decrease its misclassification rate. When sample class proportions do not reflect population proportions, the population proportion of class k within a node t may no longer be estimated by $n_k(t)/n$, but by $(n_k(t)/n) \times (n\pi_k/n_k)$ instead. Priors may be interpreted as *weights*, where each class k case would be counted $w_k \propto n\pi_k/n_k$ times.

Loss function

CART allows incorporation of variable and possibly non-symmetric misclassification costs into the splitting rule via prior specification. Intuitively, in a two-class problem, if misclassifying class 2 cases costs twice as much as misclassifying class 1 cases, then class 2 cases can be viewed as counting twice; this is in effect equivalent to having the class 2 prior twice as large as that on class 1. More precisely, let $Q(l|h)$ denote the proportion of class h cases in $\mathcal{L}$ misclassified into class l, then the resubstitution estimate for the expected misclassification cost for a tree T is

$$R(T) = \sum_{h,l} L(h,l)Q(l|h)\pi_h.$$

The overall cost $R(T)$ remains the same for altered forms π'_h and $L'(h,l)$ of π_h and $L(h,l)$ such that

$$L'(h,l)\pi'_h = L(h,l)\pi_h.$$

If for each class h, there is a constant misclassification cost $L(h,l) = L_h I(h \neq l)$, then the cost structure can be incorporated in tree building by altering the class h prior as follows:

$$\pi'_h = L_h \pi_h / \sum_l L_l \pi_l.$$

One can then proceed with the redefined class priors π'_h and a unit cost structure. In general, when costs are variable within a given class, the adjusted priors should be chosen so that the new costs $L'(h,l)$ are as close as possible to unit cost.

Variable combinations

The simplest splits are based on the value of a single variable. This results in a partition of the feature space $\mathcal{X} = \Re^G$ into rectangles, with class boundaries parallel to the axes. CART also allows for consideration of other types of variable combinations including: linear combinations and Boolean combinations (Section 5.2 of Breiman et al., 1984). Boolean splits may be beneficial when a large number of categorical explanatory variables are present. Implementation of these combinations is not straightforward and involves search algorithms that may, in some cases, lead to local optima. In a recent manuscript, Ruczinski et al. (2001) propose an adaptive regression methodology, termed *logic regression*, that attempts to construct predictor variables as Boolean combinations of binary covariates.

3.2.7 Support vector machines

Support vector machines (SVMs) were introduced in the 1970s by Vapnik (1979, 1998), but did not gain popularity until recently. SVMs are designed for binary classification (outcome coded as -1 and 1); however, they can be generalized to deal with polychotomous outcomes by considering several binary problems simultaneously. It is customary for SVMs to use the one-against-all approach (see Section 3.3.6). The following discussion of SVMs is based mainly on Burges (1998) and concerns binary classification only.

The main idea underlying SVMs is very intuitive: find the best hyperplane separating the two classes in the learning set, where *best* is described by a constrained maximization problem. Generally, one tries to maximize the so-called margin (i.e., the sum of the distances from the hyperplane to the closest positive and negative correctly classified observations, while penalizing for the number of misclassifications). One can search for the hyperplane in the original space (linear SVMs) or in a higher–dimensional space (nonlinear SVMs). The linear case is discussed first.

Linear support vector machines

Linear separable case

In the simplest application of SVMs, the two classes in the learning set can be completely separated by a hyperplane (i.e., the data are said to be *linearly separable*). This special case was first considered at the end of the 1950s by Rosenblatt (1957) and the resulting hyperplane is referred to as *perceptron*. The solution involves a linear programming problem and is described next.

Suppose a certain hyperplane separates the positive (class 1) from the negative (class -1) observations. The points $\mathbf{x}$ that lie on the hyperplane satisfy $\mathbf{w} \cdot \mathbf{x} + b = 0$, where the G-vector $\mathbf{w}$ is normal to the hyperplane and $\mathbf{w} \cdot \mathbf{x}$ denotes the inner product in $\Re^G$, i.e., scalar or dot product $\mathbf{w} \cdot \mathbf{x} = \mathbf{w}\mathbf{x}' = \sum_{g=1}^{G} w_g x_g$. Then, $|b|/||\mathbf{w}||$ is the perpendicular distance from the hyperplane to the origin, where $||\mathbf{w}|| = (\mathbf{w} \cdot \mathbf{w})^{1/2}$ is the norm of the vector $\mathbf{w}$. Define the *margin* of a separating hyperplane as the sum of the distances from the hyperplane to the closest positive and negative observations. For the linearly separable case, the separating hyperplane with the largest margin is sought. The problem can be reformulated as follows. Because the data are linearly separable, one can find a pair of hyperplanes $H_1 : \mathbf{w} \cdot \mathbf{x} + b = 1$ and $H_2 : \mathbf{w} \cdot \mathbf{x} + b = -1$ such that for all observations i, $i = 1, \ldots, n$,

$$\begin{aligned} H_1 &: \mathbf{w} \cdot \mathbf{x_i} + b \geq 1, && \text{for } y_i = 1, \\ H_2 &: \mathbf{w} \cdot \mathbf{x_i} + b \leq -1, && \text{for } y_i = -1. \end{aligned}$$

Note that the two hyperplanes are parallel, with no learning set observation falling between them. The two hyperplane inequalities can be combined into a single inequality

$$y_i(\mathbf{w} \cdot \mathbf{x_i} + b) - 1 \geq 0, \qquad \forall i. \tag{3.5}$$

Perpendicular distances of H_1 and H_2 from the origin are $|1 - b|/||\mathbf{w}||$ and $|-1 - b|/||\mathbf{w}||$, respectively. Thus, the margin is $2/||\mathbf{w}||$ and one seeks to minimize $||\mathbf{w}||^2$ subject to the constraint $y_i(\mathbf{w} \cdot \mathbf{x_i} + b) - 1 \geq 0, \forall i$. Note that the only points whose removal would affect the solution are the ones for which equality for the constraint in Equation (3.5) holds (i.e., the points lying on one of the two hyperplanes, H_1 and H_2). Those points are called *support vectors*.

It is beneficial to reformulate the problem in terms of Lagrangian multipliers. The optimization problem becomes easier to solve and the learning data only appear in the algorithm in the form of dot products between the feature vectors. The latter property allows generalization of the procedure to the nonlinear case considered below. Nonnegative Lagrangian multipliers α_i are introduced for each of the preceding constraints. It can be shown that for the linear separable case, support vector training amounts to maximizing the *dual*

$$L_D = \sum_i \alpha_i - \frac{1}{2} \sum_{i,j} \alpha_i \alpha_j y_i y_j \mathbf{x_i} \cdot \mathbf{x_j}$$

with respect to the α_i, subject to the constraints $\alpha_i \geq 0$ and $\sum_i \alpha_i y_i = 0$ for all i. The solution is given by

$$\mathbf{w} = \sum_i \alpha_i y_i \mathbf{x_i}.$$

Only the points for which $\alpha_i > 0$ are the support vectors and these points lie on one of the hyperplanes, H_1 or H_2. The removal of the remaining points would not affect the solution.

Linear nonseparable case

When the above algorithm is applied to nonseparable data, the objective function becomes arbitrarily large and there is no feasible solution. One needs to relax the constraints on the two hyperplanes while introducing a penalty for doing so. Cortes and Vapnik (1995) introduced positive *slack variables* $\xi_i \geq 0$ in the constraints which then become

$$\begin{aligned} H_1: \quad & \mathbf{w} \cdot \mathbf{x_i} + b \geq 1 - \xi_i, \qquad && \text{for } y_i = 1, \\ H_2: \quad & \mathbf{w} \cdot \mathbf{x_i} + b \leq -1 + \xi_i, \qquad && \text{for } y_i = -1. \end{aligned}$$

Note that $\sum_i \xi_i$ is an upper limit on the number of misclassifications in the learning set and the new objective function to minimize becomes

$$||\mathbf{w}||^2/2 + C(\sum_i \xi_i)^k.$$

The scalar C is a *cost* or *penalty* parameter chosen by the user, with higher C assigning a higher penalty to errors. Solving the minimization question above is a convex programming problem for any value of the parameter $k > 0$. When $k = 1$ or $k = 2$, the problem reduces to quadratic programming. SVM implementations typically set $k = 1$, so that the slack variables and their Lagrange multipliers may be ignored. It is then sufficient to solve the following quadratic programming optimization problem, which involves maximizing the dual

$$L_D = \sum_i \alpha_i - \frac{1}{2} \sum_{i,j} \alpha_i \alpha_j y_i y_j \mathbf{x_i} \cdot \mathbf{x_j}$$

subject to $0 \leq \alpha_i \leq C$ and $\sum_i \alpha_i y_i = 0$ for all i. The solution is given by

$$\mathbf{w} = \sum_{i \in S} \alpha_i y_i \mathbf{x_i},$$

where S is the set of support vectors. Hence, the only difference with the linear separable case is that now the α_i are bounded above by C.

Nonlinear support vector machines

In many situations, it is useful to consider nonlinear (in $\mathbf{x}$) decision functions or class boundaries. Mapping the data into higher dimensional spaces and then reducing the problem to the linear case leads to a simple solution. Recall again that after reformulating the optimization problem in terms of Lagrangians, the feature vectors $\mathbf{x}$ only appear in the optimization problem via dot products $\mathbf{x_i} \cdot \mathbf{x_j}$. Thus, if the data are mapped using $\Phi(\cdot)$ from, say, $\Re^G$ to $\mathcal{H}$, then the learning algorithm depends on the data only through $\Phi(\mathbf{x_i}) \cdot \Phi(\mathbf{x_j})$. If there is a *kernel function* $K(\mathbf{x_i}, \mathbf{x_j}) = \Phi(\mathbf{x_i}) \cdot \Phi(\mathbf{x_j})$, then one does not even need to know $\Phi(\cdot)$ explicitly. Widely used kernels are given in the following table.

Classifier	Kernel
Polynomial of degree p	$K(\mathbf{x}, \mathbf{y}) = (\mathbf{x} \cdot \mathbf{y} + 1)^p$
Gaussian radial basis function (RBF)	$K(\mathbf{x}, \mathbf{y}) = \exp(-\vert\vert\mathbf{x} - \mathbf{y}\vert\vert^2/2\sigma^2)$
Two-layer sigmoidal neural network	$K(\mathbf{x}, \mathbf{y}) = \arctan(\kappa \mathbf{x} \cdot \mathbf{y} - \delta)$

Note that raising the learning set into a higher and possibly infinite dimensional space (as in RBF) does not make computations prohibitively expensive. Indeed, one need not apply the mapping Φ to individual observations, but simply use the kernel K in place of the dot product of transformed observations. Mercer's condition specifies for which kernels K there exists a pair $(\mathcal{H}, \Phi)$ such that $K(\mathbf{x}, \mathbf{y}) = \Phi(\mathbf{x}) \cdot \Phi(\mathbf{y})$.

Solution methods and user-defined parameters

The support vector optimization problem can be solved analytically only when the size of the learning set is very small or for separable cases where it is known which of the observations become support vectors. In general, the problem must be solved numerically. While traditional optimization methods such as Newton and quasi-Newton algorithms are not directly applicable for technical reasons, decomposition methods, that use the above algorithms on appropriately selected subsets of the data, may be applied. The `LIBSVM` software interfaced with R uses such an approach (Chang and Lin, 2001).

Here, we consider the *C-classification* method for SVMs. The user-specified parameters for this method are the kernel K, i.e., the nonlinear transformation of the original data, and the cost parameter C for misclassifications when the learning data are non-separable in the transformed space. Other types of SVMs include: ν-classification, where the parameter ν provides an upper bound on the fraction of misclassified learning set observations and a lower bound on the fraction of support vectors (Chang and Lin, 2001). Perhaps the largest limitation of SVMs lies in the difficulty for the user to choose an appropriate kernel for a given problem.

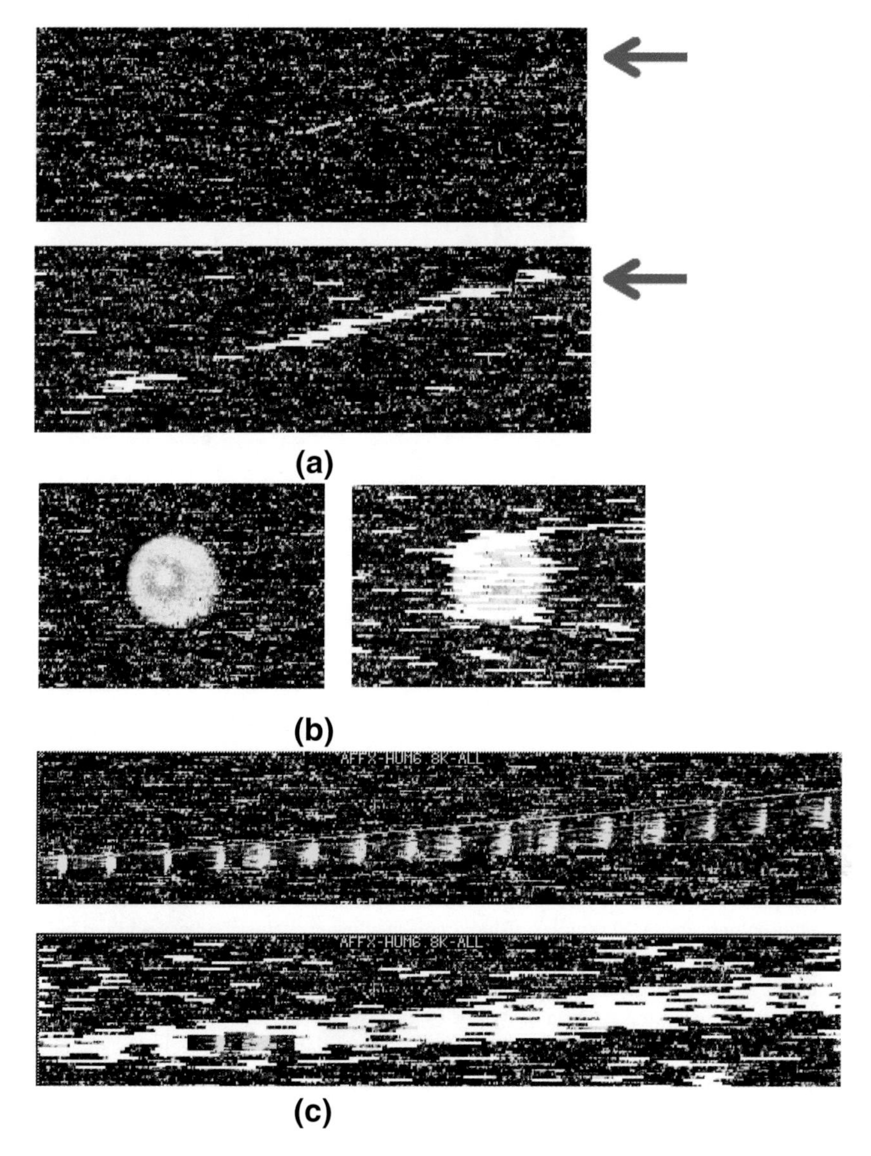

FIGURE 1.6 (a) A long scratch contamination (indicated by arrow) is alleviated by automatic outlier exclusion along this scratch. (b, c) Regional clustering of array-outliers (white bars) indicate contaminated regions in the original images. These outliers are automatically detected and accommodated in the analysis. Note that some probe sets in the contaminated region are not marked as array-outliers, because the contamination contributes additively to PM and MM in a similar magnitude and thus cancels in the PM–MM differences, preserving the correct signals and probe response patterns. (From Li, C. and W. H. Wong, *Proc. Natl. Acad. Sci.*, 98:31–36, 2001a. With permission.)

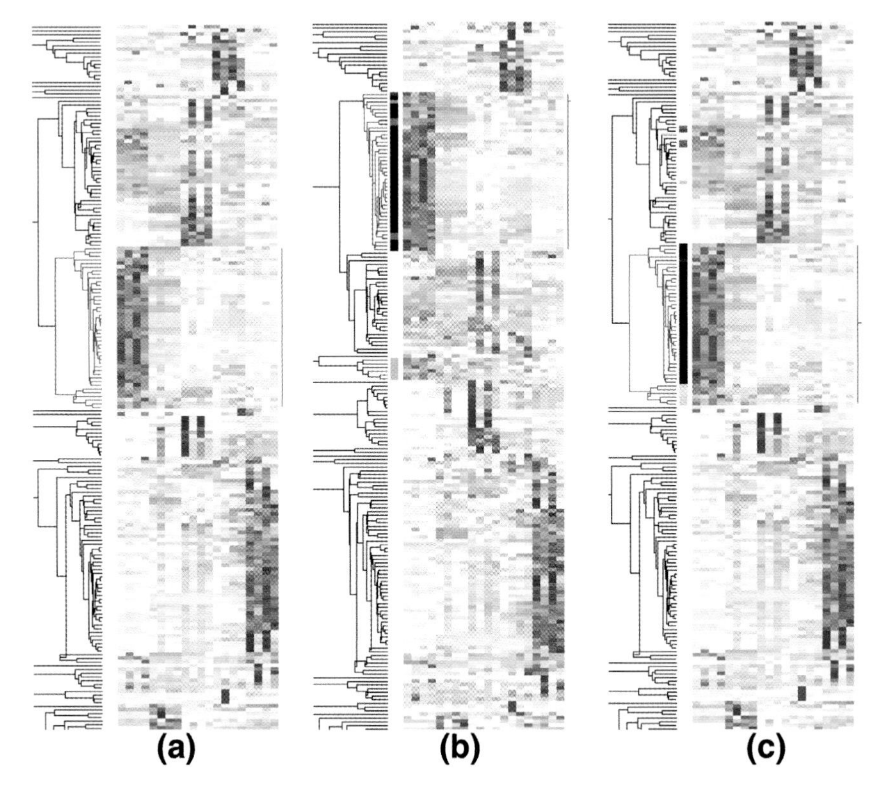

FIGURE 1.15 (a) 225 filtered genes are clustered based on their expression profiles across the 20 samples. Each gene's expressions values are standardized to have mean zero and standard deviation one across 20 samples. Blue represents lower expression level and red higher expression level. We may be particularly interested in the gene branch colored blue. (b) The clustering tree after a particular resampling. Although the original "blue" genes are scattered to various places, using the critieria described in the text, we can still determine the location of the original cluster. (c) After resampling 30 times, the reliability of the genes belonging to the original cluster is indicated by the vertical gray-scale bar on the right of the clustering tree. (From Li, C. and W. H. Wong, *Genome Biol.*, 2(8):0032.1–0032.11, 2001b. With permission.)

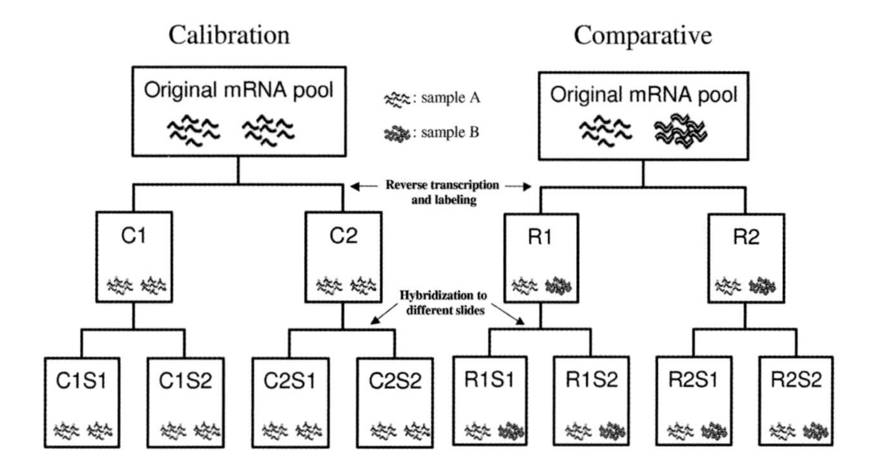

FIGURE 1.16 In calibration experiments, same samples are applied in the two dyes whereas, in comparative experiments, different samples are used with the two dyes. Replications are performed in experiment and slide stage. The 125-gene project has a similar design except for the quadruple spotting.

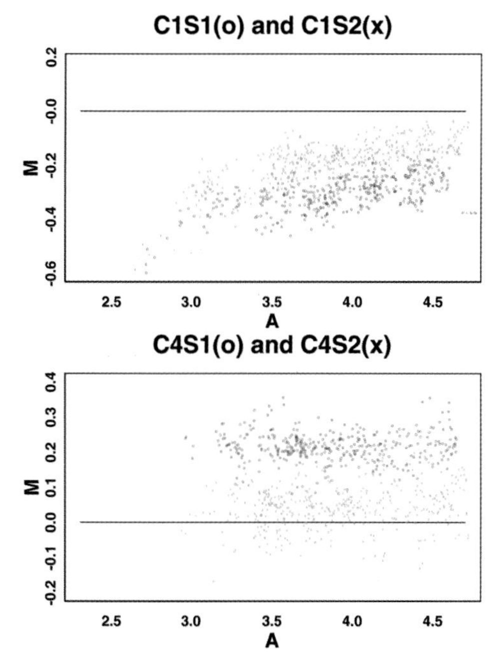

FIGURE 1.19 MA-plots of two slides in the same calibration experiment. The upper plot illustrates a different pattern of MA-plot on C1S1(o) and C1S2(x) in the 125-gene project. The lower MA-plot in calibration 4 also illustrates the same situation. Thus, the normalization curve is slide-dependent and should be estimated and applied within the same slide. (From Tseng, G. C. et al., *Nucleic Acids Res.*, 29(12):2549–2557, 2001. With permission.)

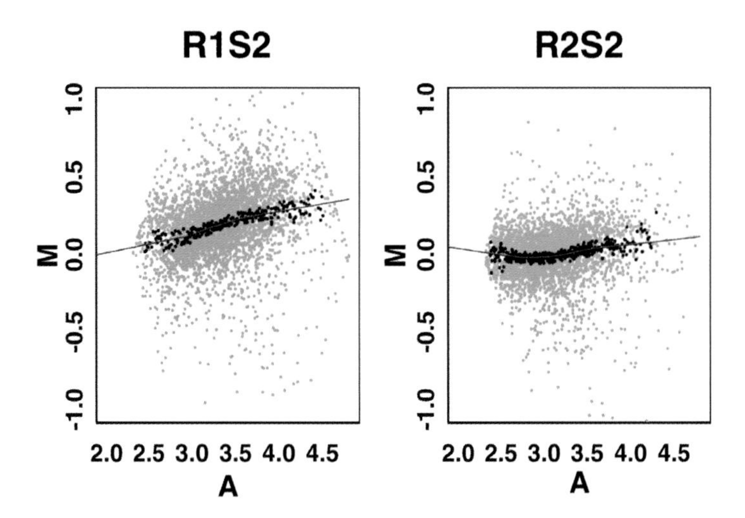

FIGURE 1.20 Normalization curves and MA-plots in comparative experiments in 4129-gene project. The darker points are genes of the rank-invariant set selected in an iterative manner (p = 0.02). (From Tseng, G. C. et al., *Nucleic Acids Res.*, 29(12):2549–2557, 2001. With permission.)

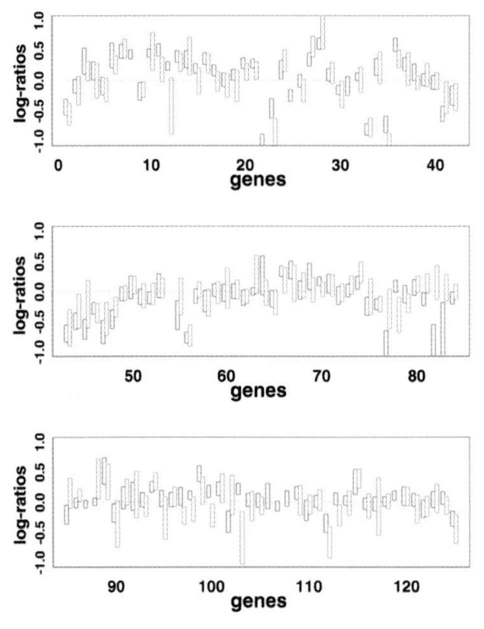

FIGURE 1.21 The orange and green rectangles show the 95% posterior interval for the underlying expression level θ_9 of the 125-gene and 4129-gene projects (green: 125-gene project; orange: 4129-gene project). Rectangles of gene 54 (aceA) are below −1.0 and do not appear in the graph. (From Tseng, G. C. et al., *Nucleic Acids Res.*, 29(12):2549–2557, 2001. With permission.)

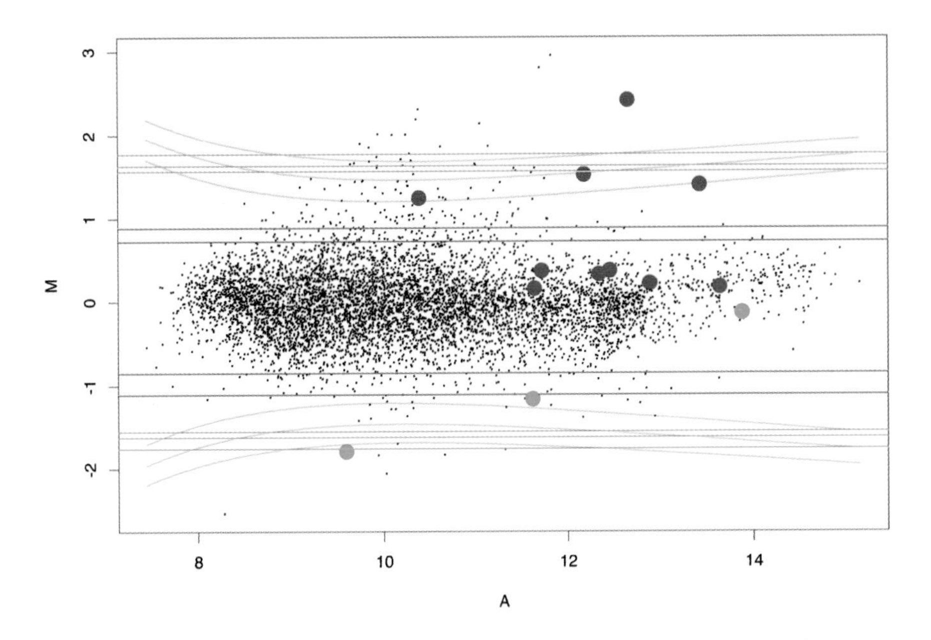

FIGURE 2.5 Single-slide methods: an MA-plot showing the contours for the methods of Newton et al., 2001 (orange, odds of change of 1:1, 10:1, and 100:1), Chen et al., 1997 (purple, 95% and 99% confidence), and Sapir and Churchill, 2000 (cyan, 90%, 95%, and 99% posterior probability of differential expression). The points corresponding to genes with adjusted p-value less than 0.05 (based on data from 16 slides) are colored in green (negative t-statistic) and red (positive t-statistic). The data are from transgenic mouse 8.

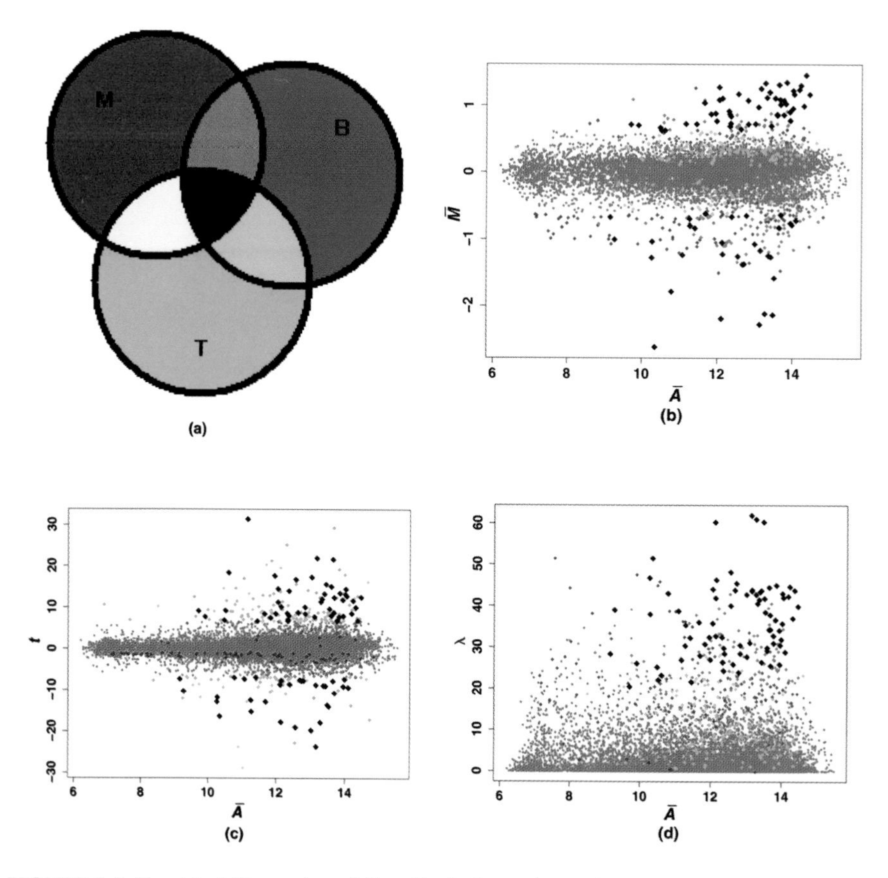

FIGURE 2.6 Graphical illustration of Case Study I: *swirl* experiment. (a) The color code that defines three groups of 250 genes consisting of largest values of $|\overline{M}|$, $|t|$, and B values; (b) $\overline{M}\,\overline{A}$-plot; (c) t vs. $\overline{A}$; (d) λ vs. $\overline{A}$.

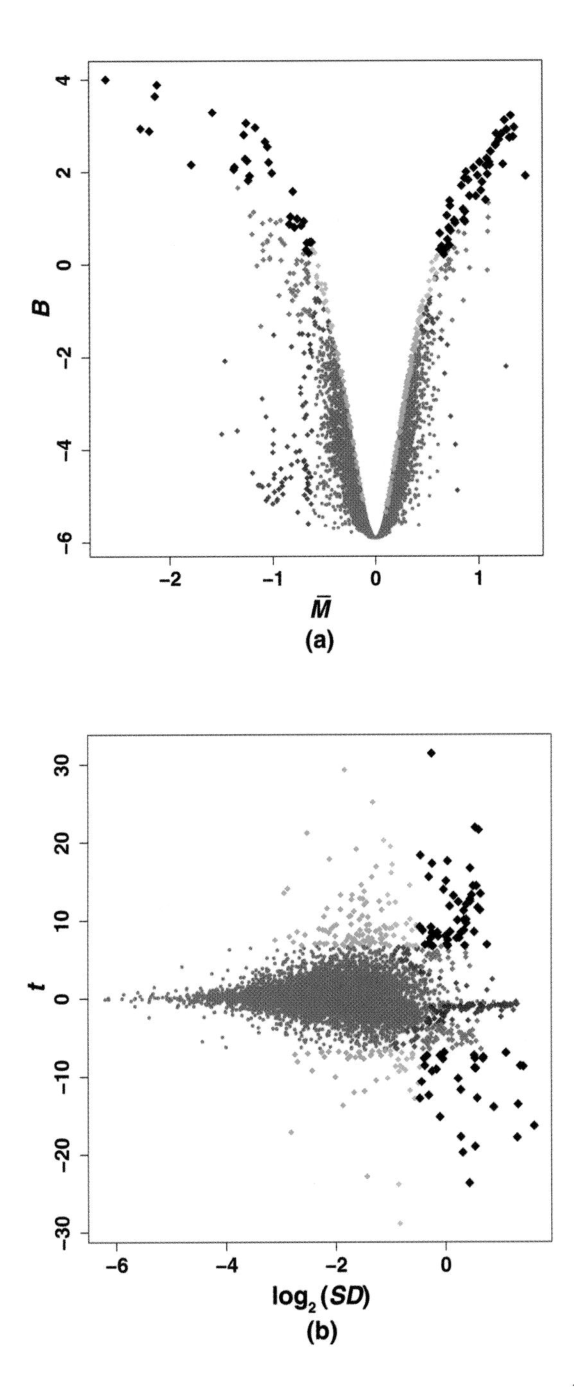

FIGURE 2.7 Case Study I: *swirl* experiment. Plots of (a) log odds B vs. $\overline{M}$ and (b) t vs. $\log_2$ (SD). Spots (genes) corresponding to large $|\overline{M}|$, $|t|$, and B values highlighted according to color code shown in Figure 2.6a.

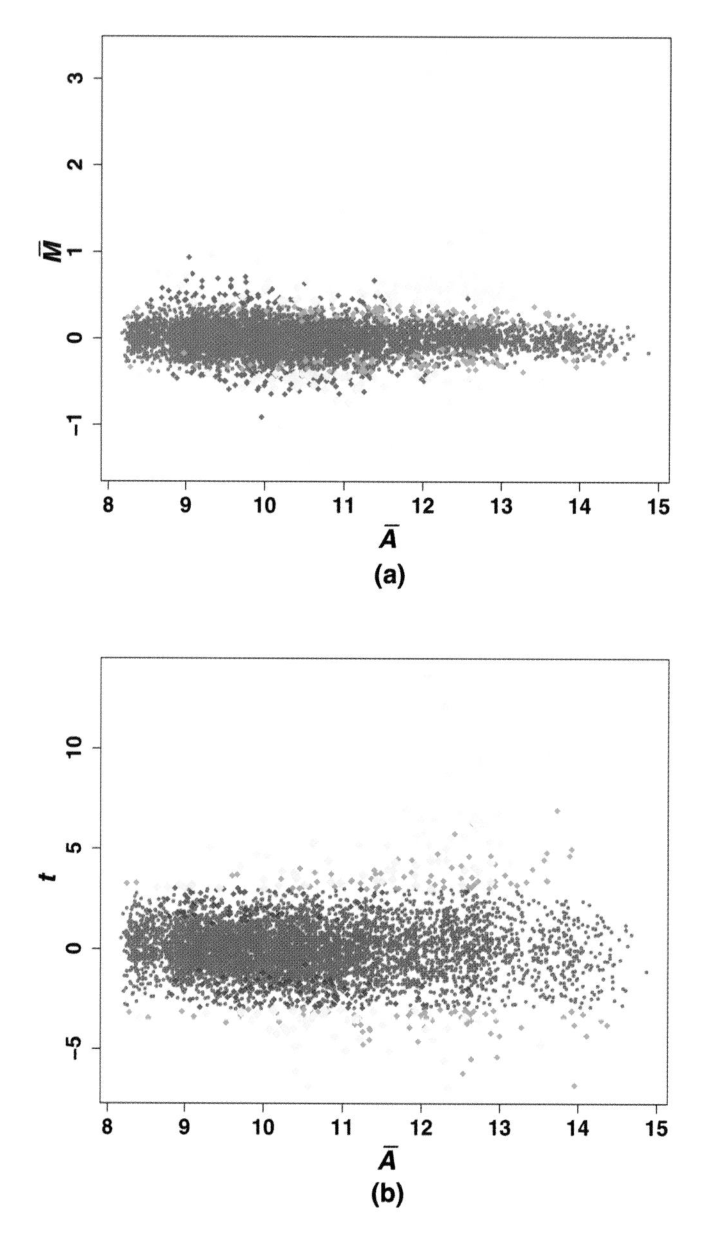

FIGURE 2.8 Case Study II SR–BI experiment: (a) $\overline{M}\,\overline{A}$-plot; (b) t vs. A. Spots (genes) corresponding to large $|\overline{M}|$ and $|t|$ values highlighted according to color code shown in Figure 2.6a.

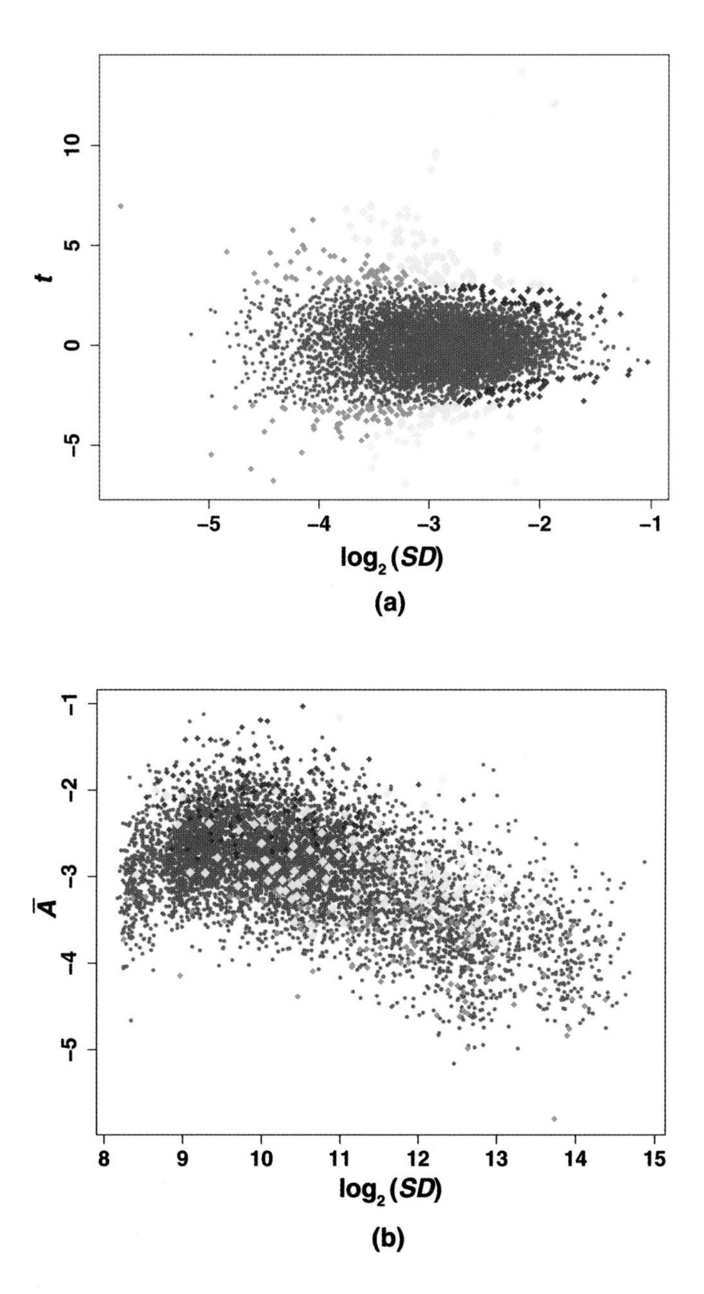

FIGURE 2.9 Case Study II: SR–BI experiment. (a) t vs. $\log_2(SD)$; (b) $\log_2(SD)$ vs. $\overline{A}$.

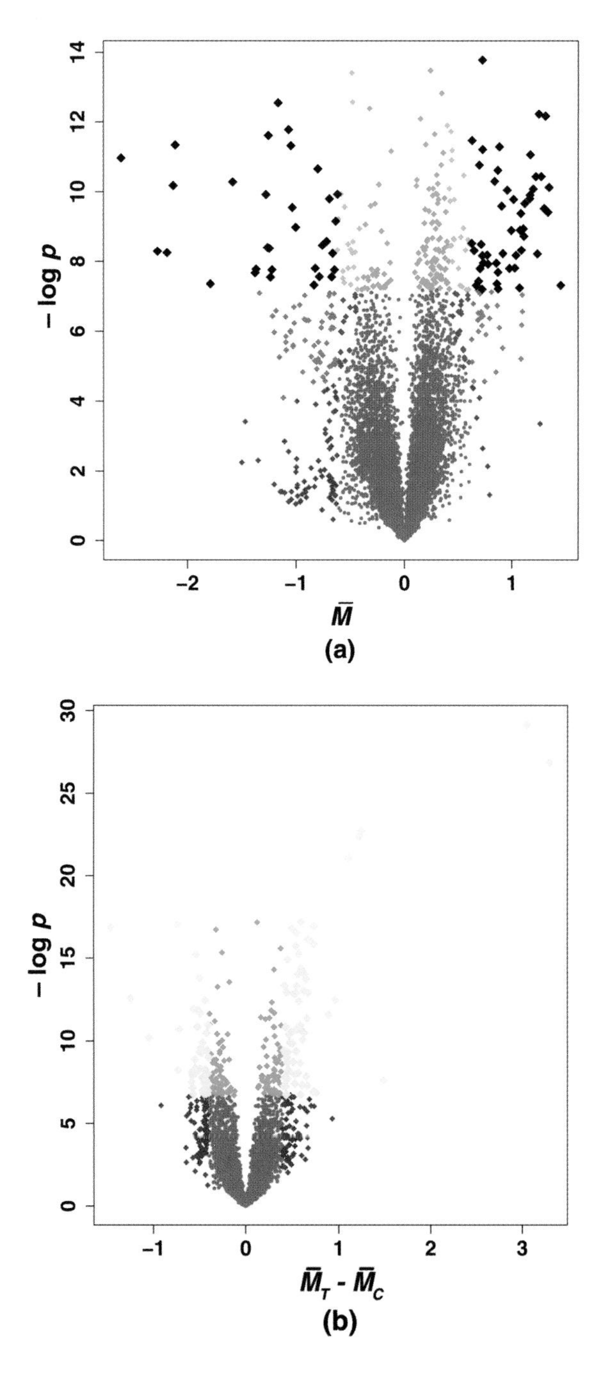

FIGURE 2.11 Case Study I: swirl experiment: – log p-value vs. average log-fold change ($\overline{M}$). (b) Case Study II: SR–BI experiment: – log p-value vs. difference of log fold change $\overline{M}_T$–$\overline{M}_C$. Color code as in Figure 2.6a.

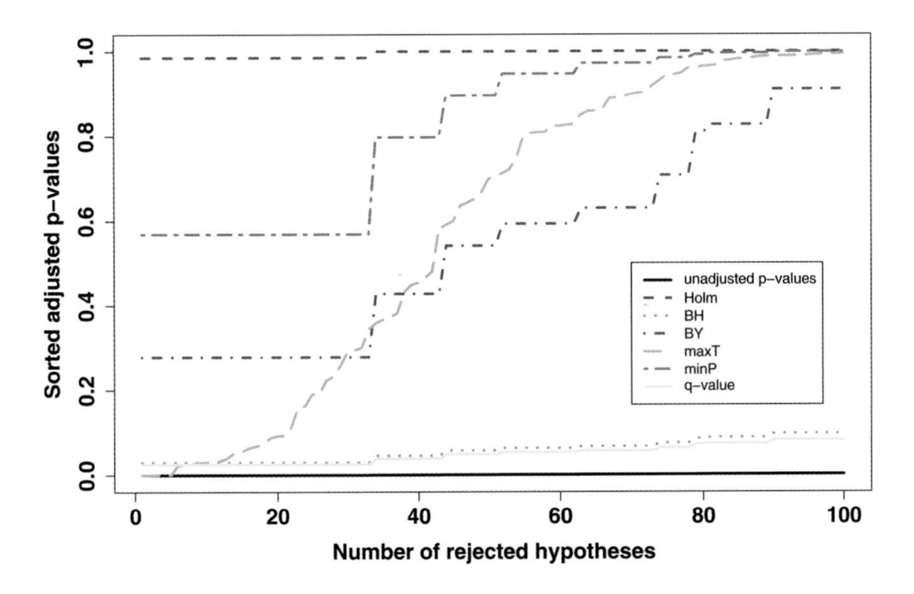

FIGURE 2.12 Comparisons of multiple testing procedures. The plots concern the spots with the 100 smallest unadjusted p-values, scored 1 to 100 horizontally. The different adjusted p-values and the q-values are vertical; see insert for line code.

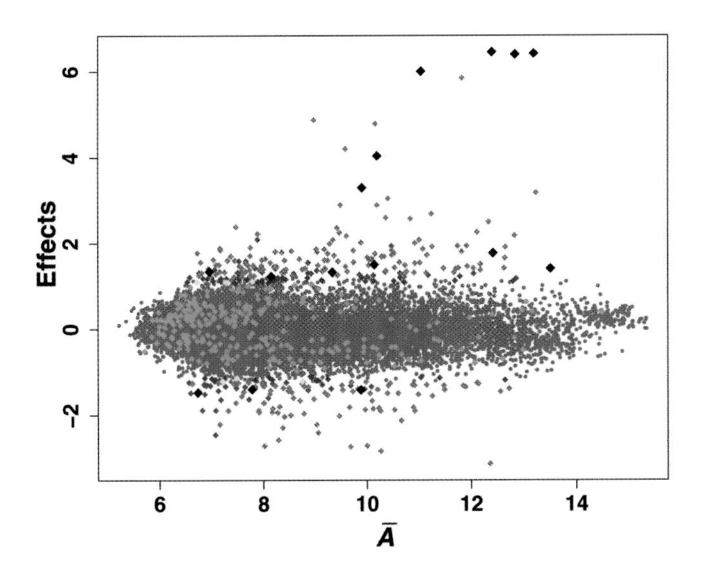

FIGURE 2.20 Estimated effect $\hat{v.t}$ vs. $\overline{A}$ plot for the interaction $v.t$ between *time* and *mutant*.

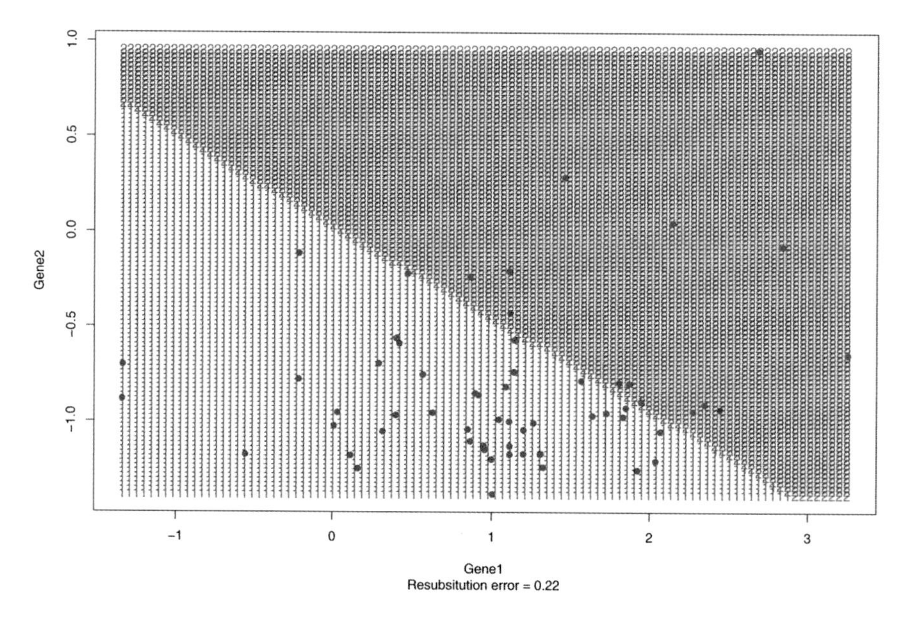

FIGURE 3.1 Brain tumor MD survival dataset, LDA. Partition produced by linear discriminant analysis (LDA) applied to the two genes with the largest absolute t-statistics. Predicted responses "survivor" and "nonsurvivor" are indicated by shades of red and blue, respectively. The entire learning set of $n = 60$ samples was used to build the classifier; the resubstitution error rate is shown below the plot. Learning set observations are plotted individually using the color for their true class.

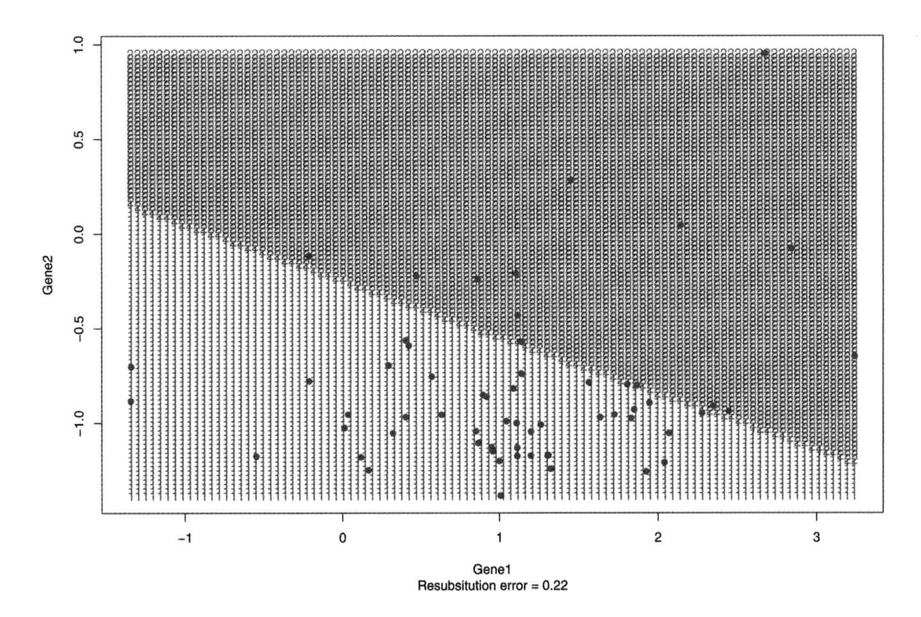

FIGURE 3.2 Brain tumor MD survival dataset, QDA. Partition produced by quadratic discriminant analysis (QDA) applied to the two genes with the largest absolute t-statistics. Predicted responses "survivor" and "nonsurvivor" are indicated by shades of red and blue, respectively. The entire learning set of $n = 60$ samples was used to build the classifier; the resubstitution error rate is shown below the plot. Learning set observations are plotted individually using the color for their true class.

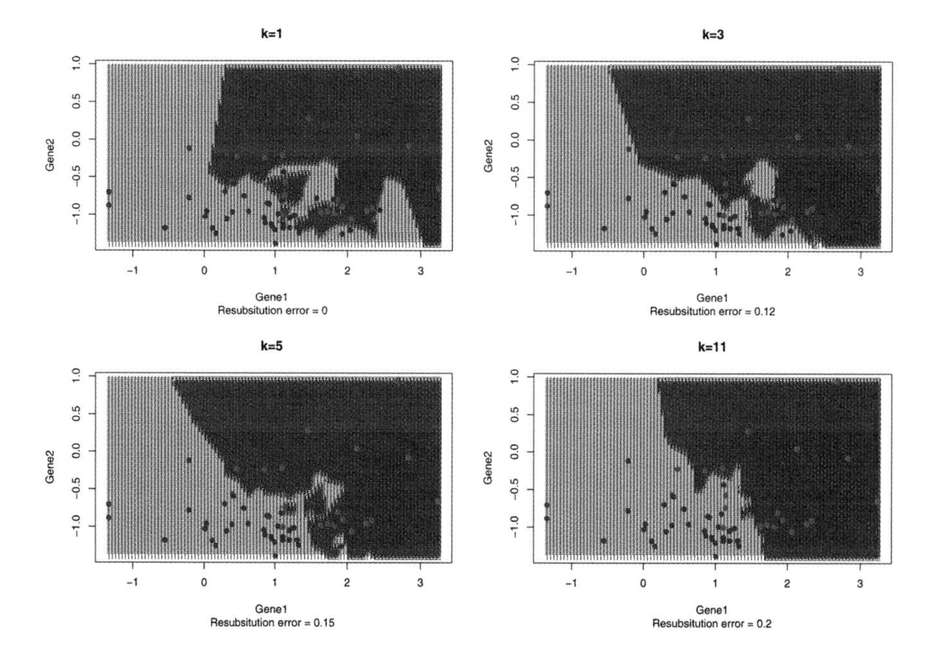

FIGURE 3.3 Brain tumor MD survival dataset, k-NN. Partitions produced by k-nearest neighbor classification ($k = 1, 3, 5, 11$) applied to the two genes with the largest absolute t-statistics. Predicted responses "survivor" and "nonsurvivor" are indicated by shades of red and blue, respectively. The entire learning set of $n = 60$ samples was used to build the classifier; the resubstitution error rate is shown below the plot. Learning set observations are plotted individually using the color for their true class.

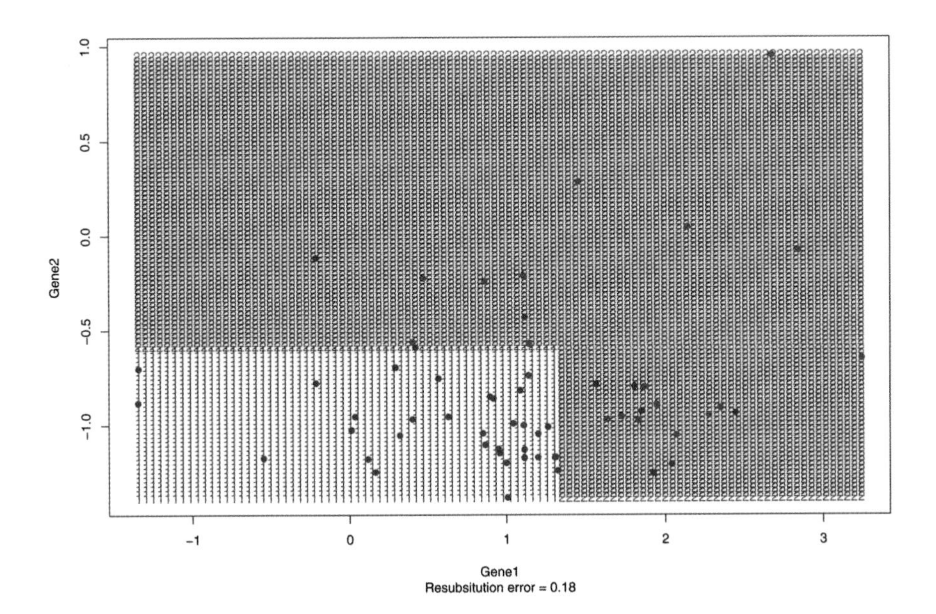

FIGURE 3.4 Brain tumor MD survival dataset, CART. Partition produced by CART (tenfold CV) applied to the two genes with the largest absolute t-statistics. Predicted responses "survivor" and "nonsurvivor" are indicated by shades of red and blue, respectively. The entire learning set of $n = 60$ samples was used to build the classifier; the resubstitution error rate is shown below the plot. Learning set observations are plotted individually using the color for their true class.

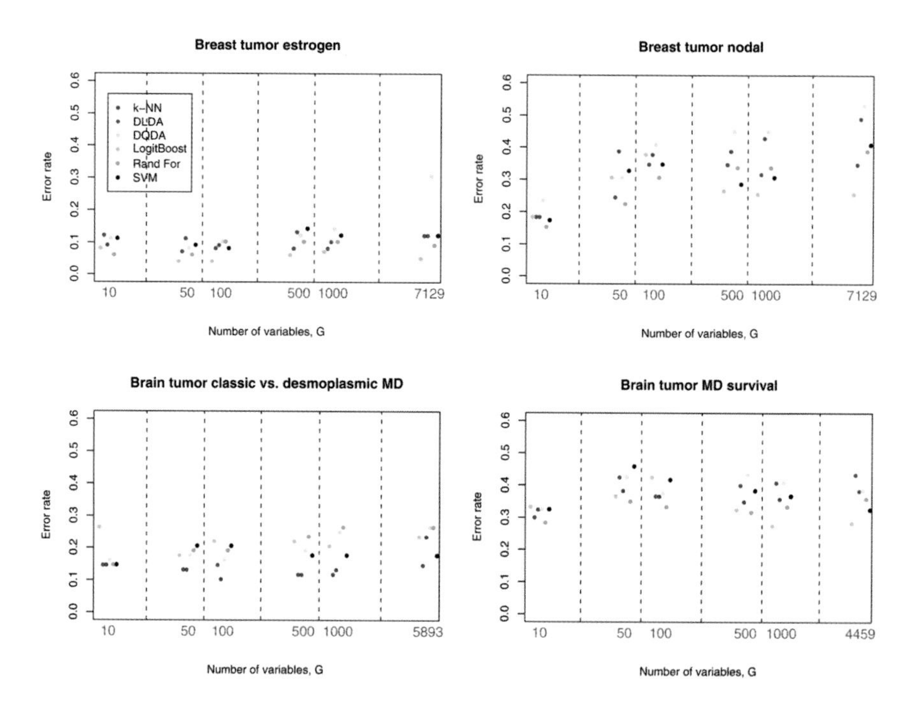

FIGURE 3.5 Comparison of classification error rates. Classification error rates were estimated for each classifier using leave-one-out cross-validation as in Table 3.3 to Table 3.6. For each family of classifiers and given number of genes G, median error rates were computed over the parameter values considered in Table 3.3 to Table 3.6. The median error rates are plotted vs. the number of genes G (on log-scales) for each family of classifiers.

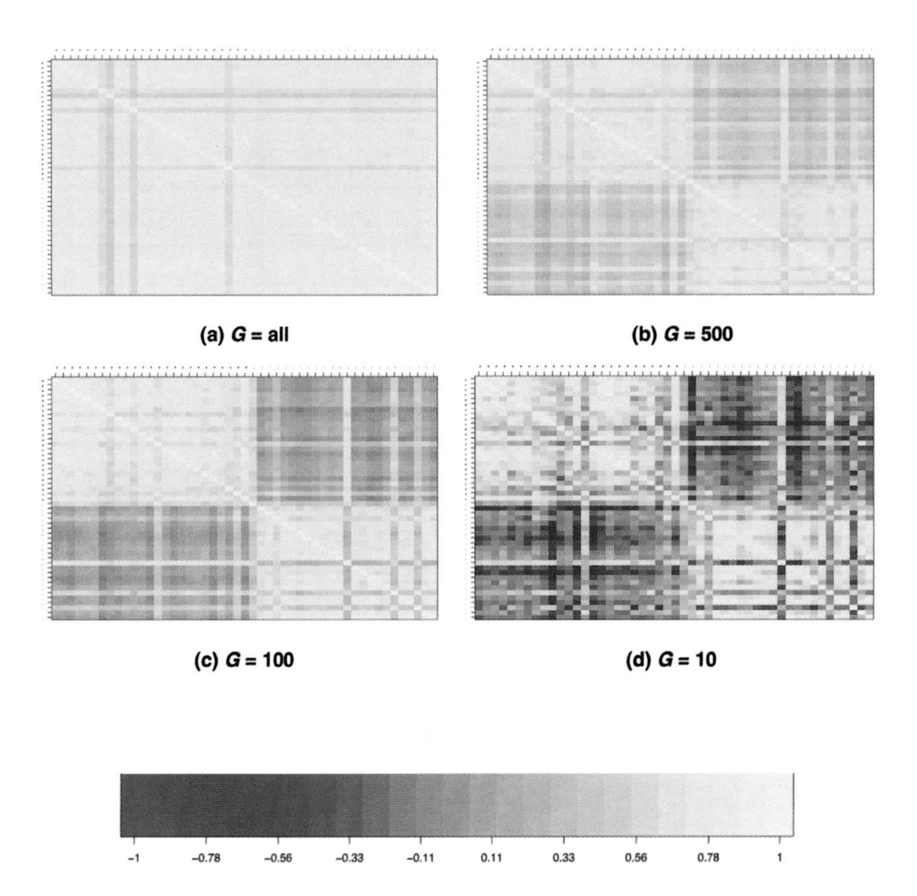

FIGURE 3.6 Breast tumor estrogen dataset. Images of the correlation matrix for the 49 breast tumor mRNA samples based on expression profiles for: (a) all $G = 7129$ genes; (b) the $G = 500$ genes with the largest absolute t-statistics; (c) the $G = 100$ genes with the largest absolute t-statistics; (d) the $G = 10$ genes with the largest absolute t-statistics. The mRNA samples are ordered by class: first ER positive, then ER negative. Increasingly positive correlations are represented with yellows of increasing intensity, and increasingly negative correlations are represented with blues of increasing intensity. The color bar below the images may be used for calibration purposes.

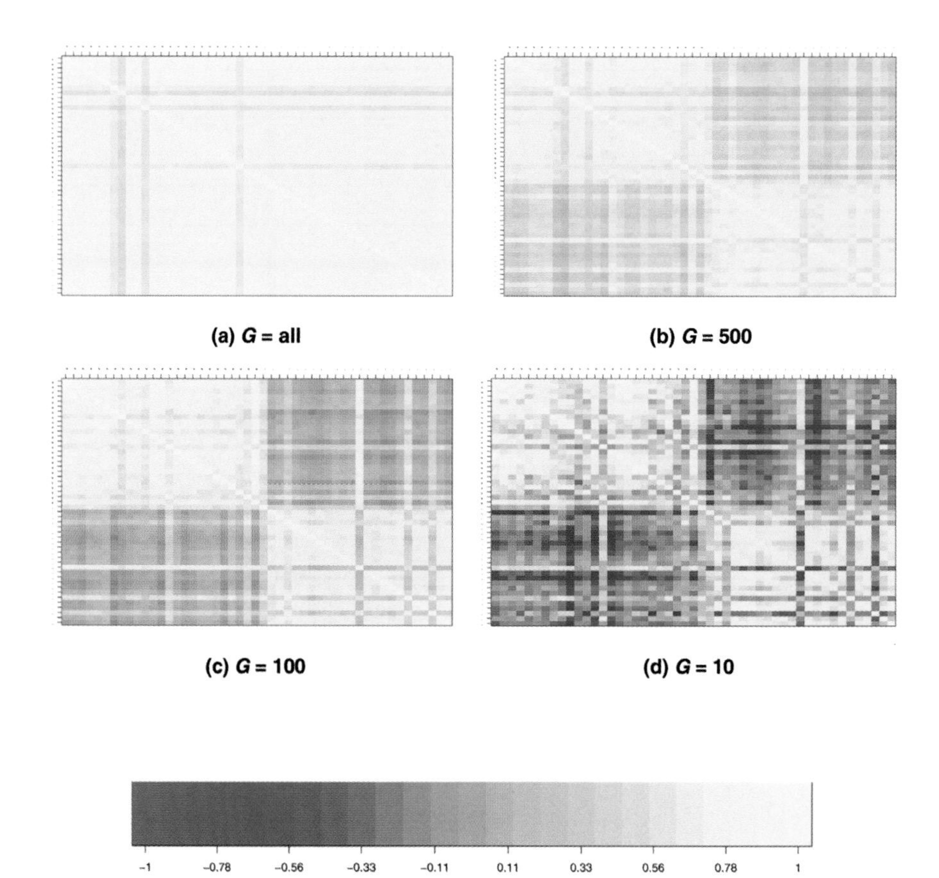

FIGURE 3.7 Breast tumor nodal dataset. Images of the correlation matrix for the 49 breast tumor mRNA samples based on expression profiles for: (a) all $G = 7129$ genes; (b) the $G = 500$ genes with the largest absolute t-statistics; (c) the $G = 100$ genes with the largest absolute t-statistics; (d) the $G = 10$ genes with the largest absolute t-statistics. The mRNA samples are ordered by class: first nodal positive, then nodal negative. Increasingly positive correlations are represented with yellows of increasing intensity, and increasingly negative correlations are represented with blues of increasing intensity. The color bar below the images may be used for calibration purposes.

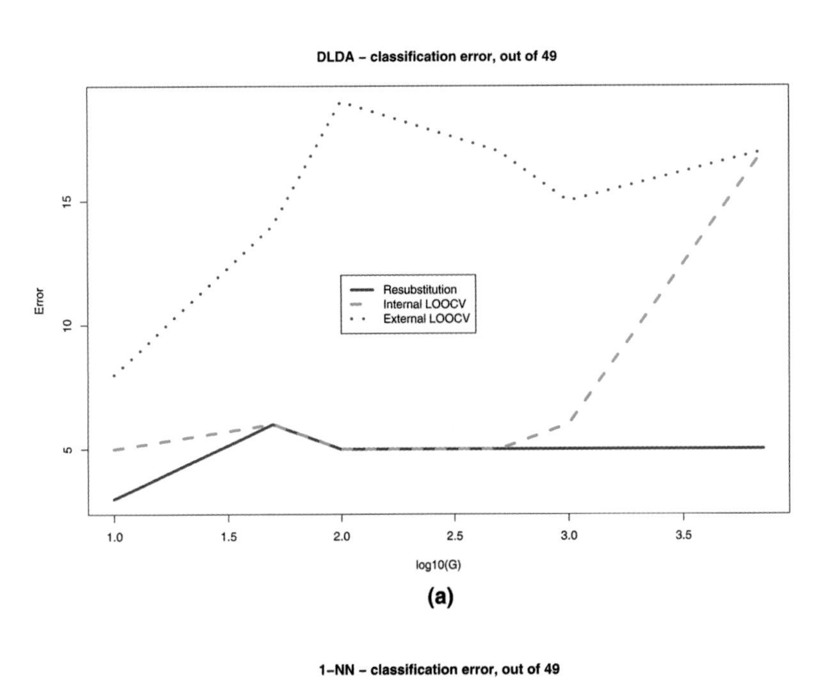

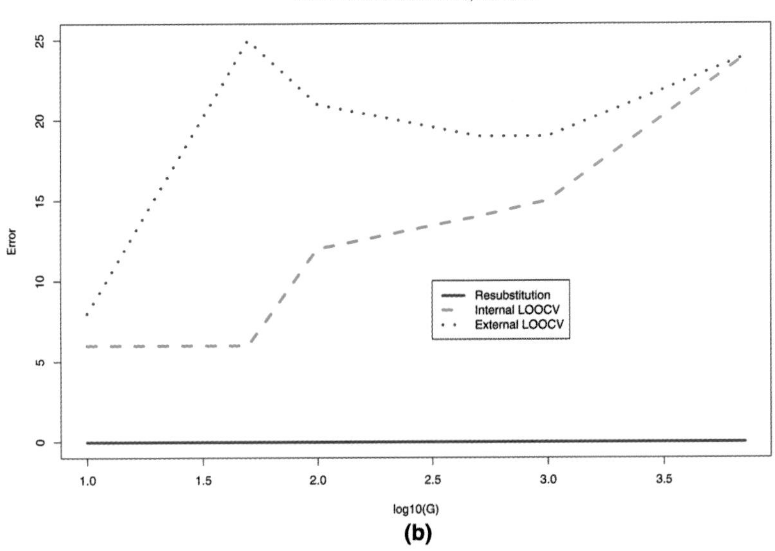

FIGURE 3.13 Breast tumor nodal dataset. Comparison of classification error estimation of procedures: (solid lines) *resubstitution estimation*, where the entire learning set is used to perform feature selection, build the classifier, and estimate classification error; (dashed lines) *internal LOOCV*, where feature selection is done on the entire learning set and LOOCV is applied only to the classifier building process; (dotted lines) *external LOOCV*, where LOOCV is applied to the feature selection *and* the classifier building process. Error rates are plotted vs. the number of genes G (on log-scale) for (a) DLDA and (b) 1-NN.

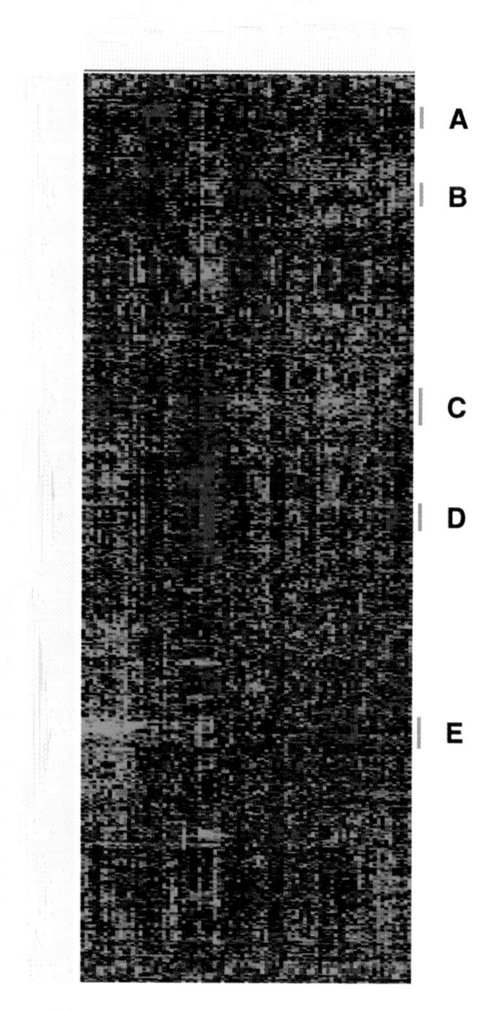

FIGURE 4.1 Heat map of the microarray data, with rows and columns arranged according to a hierarchical clustering method. Grey pixels represent missing data. Letters A–E and corresponding grey bars represent groups of genes displayed in greater detail in Figure 4.2.

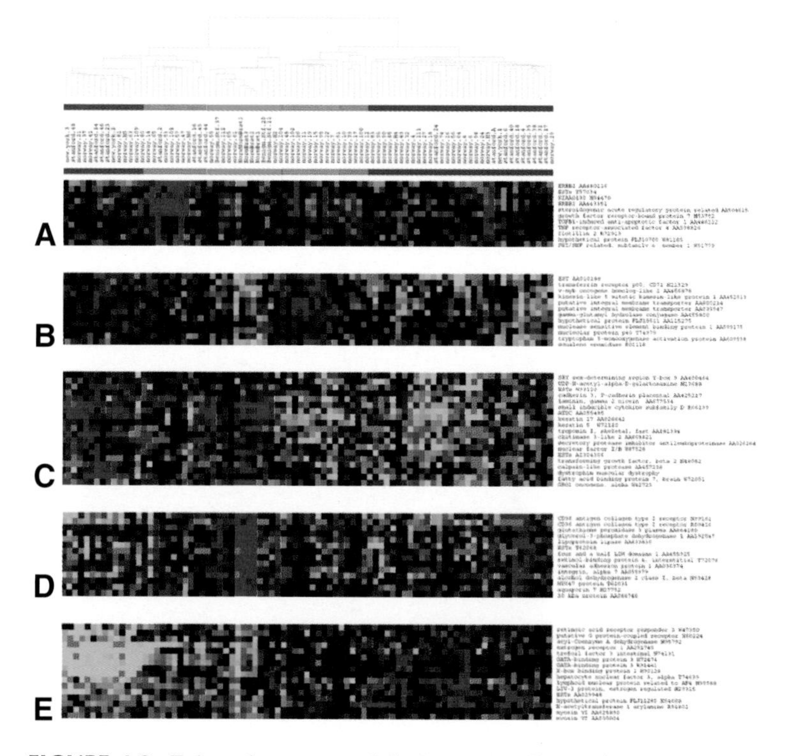

FIGURE 4.2 Enlarged segments of the heat map. Interesting groups of genes (rows) have been selected from Figure 4.1, as indicated by the letters A–E. Tumor subtypes corresponding to clusters of columns are indicated by color bars on the upper dendogram. See text for discussion.

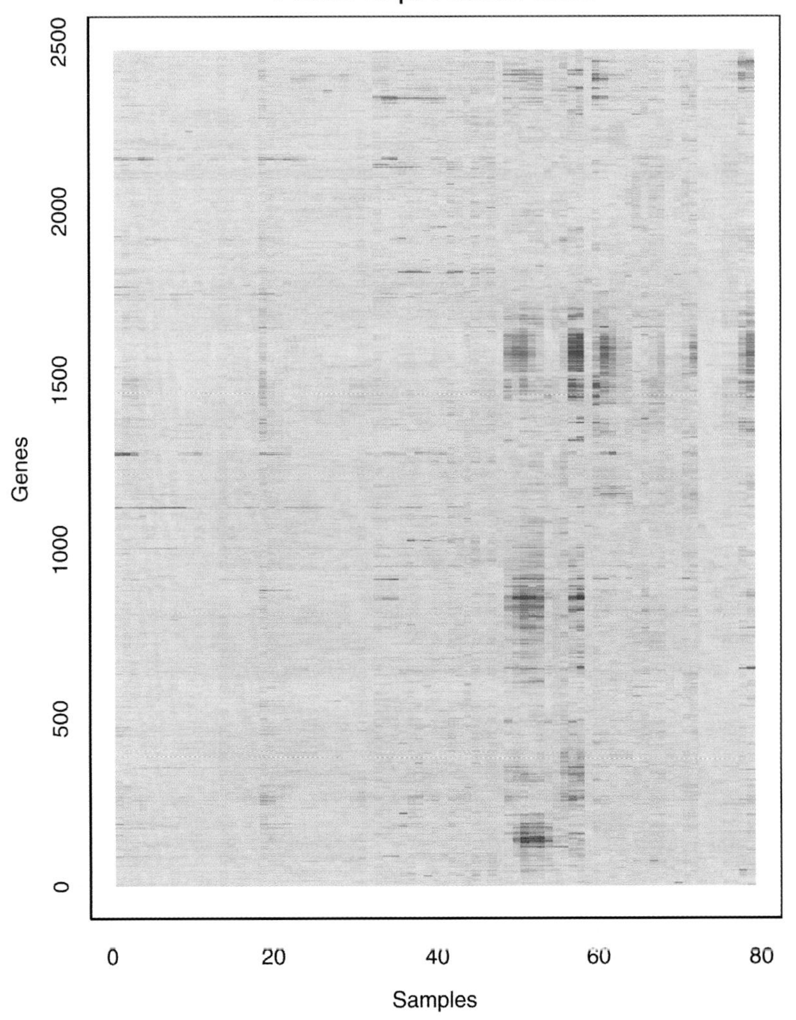

FIGURE 4.21 Yeast expression data.

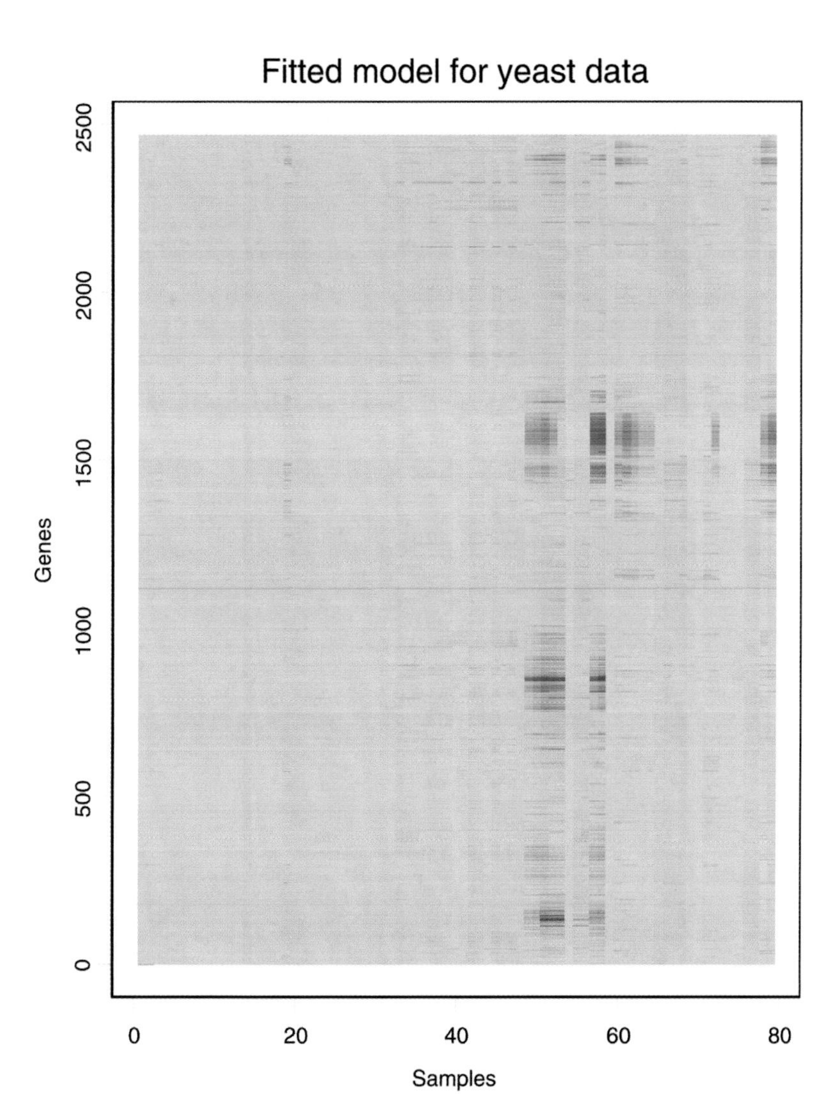

FIGURE 4.22 Fitted plaid model for yeast expression data.

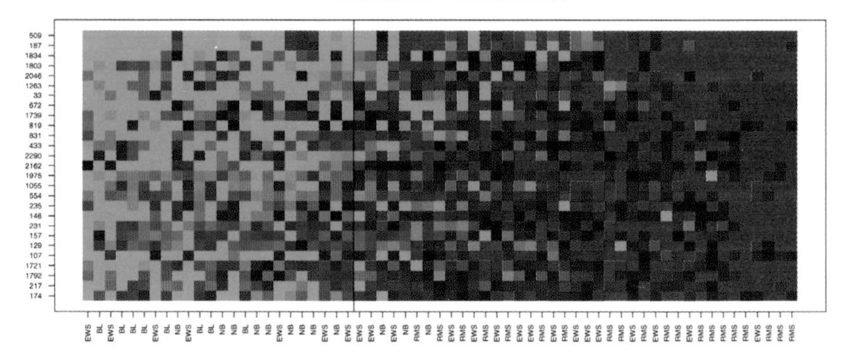

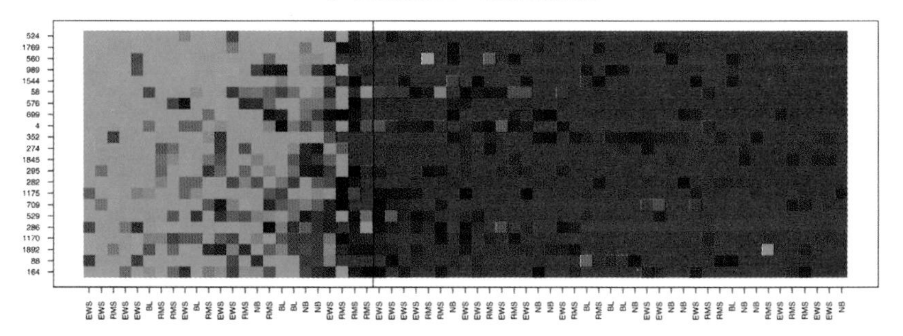

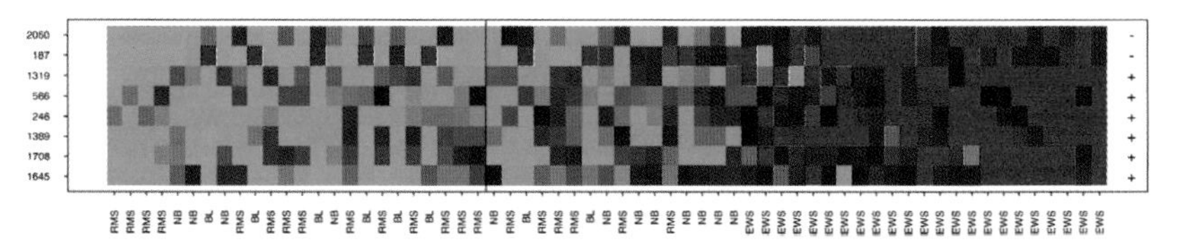

FIGURE 4.27 The first three shaves for the SRBCT cancer data. In each panel, we see the cluster of genes ordered according to their supergene (their signed average); columns (samples) are ordered by mean expression within each shave; the cancer subclass labels are indicated below each panel. The + and – signs in the third shave indicate that some genes had their signs flipped in the averaging.

3.3 General issues in classification

This section discusses the following important topics in classification: feature selection, standardization, distance function, loss function, class representation, imputation of missing data, and polychotomous classification. Decisions regarding these issues can be made implicitly by choosing a certain classifier (e.g., distance function in LDA) or explicitly (e.g., distance function in nearest neighbor classifiers). In addition, the impact of these decisions depends on the classifier (e.g., trees tend to be less sensitive to the feature selection scheme than nearest neighbor predictors). However, all these choices are part of the classifier training process and should be taken into account in performance assessment as described in Section 3.4.

3.3.1 Feature selection

Feature selection is one of the most important issues in classification; it is particularly relevant in the context of microarray datasets with thousands of features, most of which are likely to be uninformative. Some classifiers like CART perform automatic feature selection and are relatively insensitive to the variable selection scheme. In contrast, standard LDA and nearest neighbor classifiers do not perform feature selection; all variables, whether relevant or not, are used in building the classifier. Current implementations of SVMs do not include automatic feature selection, and there is some evidence that they do not perform well in the presence of a large number of extraneous variables (Weston et al., 2001). For many classifiers, it is thus important to perform some type of feature selection, otherwise performance could degrade substantially with a large number of irrelevant features. Feature selection may be performed *explicitly*, prior to building the classifier, or *implicitly*, as an inherent part of the classifier building procedure, for example, using modified distance functions (see Section 3.3.2). In the machine learning literature, these two approaches are referred to as *filter* and *wrapper* methods, respectively.

Filter methods

The simplest gene screening procedures are *one-gene-at-a-time* approaches, in which genes are ranked based on the value of univariate test statistics, such as: t- or F-statistics (Dudoit et al., 2002a); ad hoc signal-to-noise statistics (Golub et al., 1999; Pomeroy et al., 2002); nonparametric Wilcoxon statistics (Dettling and Bülmann, 2002; Park et al., 2001); p-values (Dudoit et al., 2002b). Possible meta-parameters for feature selection include the number of genes G or a p-value cut-off. A formal choice of these parameters may be achieved by cross-validation or bootstrap procedures.

More refined feature selection methods consider the *joint* distribution of the gene expression measures. In a recent article, Bø and Jonassen (2002) investigate possible advantages of subset selection procedures for screening gene pairs to be used in classification. The authors show that bivariate approaches achieve better results with fewer variables compared to univariate ones. Other methods include ordering variables according to their *importance* as defined in random forests (see Breiman, 1999 and Section 3.5.3 in this book). There, all the features are considered simultaneously, hence allowing the detection of genes with weak main effects but strong interactions.

Wrapper methods

Feature selection may also be performed implicitly by the classification rule itself. In this case, different approaches to feature selection will be used by different classifiers. In classification trees (e.g., CART, described in Section 3.2.6), features are selected at each step based on reduction in impurity and the number of features used (or size of the tree) is determined by pruning the tree using cross-validation. Thus, feature selection is an inherent part of tree building and pruning deals with the issue of over-fitting. The nearest shrunken centroid method is an extension of DLDA that performs automatic feature selection by shrinking class mean vectors toward the overall mean vector based on the value of t-like statistics (Section 3.2.4 in this book and Tibshirani et al., 2002). Suitable modifications of the distance function in nearest neighbor classification allow automatic feature selection (Section 3.2.5). In the context of Bayesian regression models, West et al. (2001) use singular value decomposition of the expression matrix X to derive *supergenes*, where the expression measures of the supergenes are linear combinations of the expression measures of individual genes.

The importance of taking feature selection into account when assessing the performance of a classifier cannot be stressed enough (Ambroise and McLachlan, 2002; West et al., 2001). Feature selection *is* an aspect of classifier training, whether done explicitly or implicitly. Thus, when using for example cross-validation to estimate generalization error, feature selection should be done *not* on the entire learning set, but separately for each cross-validation sample used to build the classifier. Leaving out feature selection from cross-validation or other resampling-based performance assessment methods results in overly optimistic error rates (see Section 3.4).

3.3.2 Standardization and distance function

Standardizing observations and variables

The transformation of variables and/or observations (e.g., location and scale transformations) is an important issue in classification, as the choice of transformation can affect the performance of the classifier. As noted next, the distance function and its behavior in the context of classification are intimately related to the scale on which measurements are made. The choice of a transformation and distance function should thus be made jointly and in conjunction with the choice of a classifier. A common type of transformation for continuous variables is *standardization*, so that each variable has mean zero and unit variance across observations. For example, in the case of gene expression data, one might consider the following standardization of the gene expression measures

$$x^*_{gi} = \frac{x_{gi} - \bar{x}_{g.}}{s_g}$$

where $\bar{x}_{g.}$ and s_g denote respectively the average and standard deviation of gene g's expression measures across the n arrays. Such a standardization in some sense puts all genes on an equal footing and weighs them equally in the classification. One could also use robust estimators of location and scale, like the median and median absolute deviation (MAD).

In principle, one could envisage standardizing either variables or observations (i.e., rows or columns of the gene expression data matrix X). In many cases, however, variables are not directly comparable and cannot be meaningfully averaged or combined within observations. In such situations, only the issue of variable standardization arises (i.e., standardization of the rows). Consider, for example, a height variable measured in meters and a weight variable measured in kg for 100 persons (observations). One may standardize both height and weight variables, so that they have mean zero and variance one across individuals (and are unit-less). Combining the two raw height and weight variables within individuals would not be meaningful before this standardization as the measurement units of m and kg are not comparable.

Standardization for microarray data

In microarray experiments, however, one could consider standardizing features (genes) or observations (arrays), as the expression measures of each gene are fluorescence intensities which are directly comparable. Standardization of observations or arrays can be viewed as part of the normalization step. In the context of microarray experiments, *normalization* refers to identifying and removing the effects of systematic variation, other than differential expression, in the measured fluorescence intensities (e.g., different labeling efficiencies and scanning properties of the Cy3 and Cy5 dyes; different scanning parameters, such as PMT settings; print-tip, spatial, or plate effects). It is necessary to normalize the fluorescence intensities before any analysis that involves comparing expression measures within or between arrays, in order to ensure that observed differences in fluorescence intensities are indeed reflecting differential gene expression, and not some printing, hybridization, or scanning artifact. Standardizing the microarray data so that arrays have mean zero and variance one achieves a location and scale normalization of the different arrays. In a study of normalization methods, we have found scale adjustment to be desirable in some cases, in order to prevent the expression measures in one particular array from dominating the average expression measures across arrays (Yang et al., 2001). Furthermore, this standardization is consistent with the common practice in microarray experiments of using the correlation between the gene expression profiles of two mRNA samples to measure their similarity (Alizadeh et al., 2000; Perou et al., 1999; Ross et al., 2000). In practice, however, we recommend general adaptive and robust normalization procedures which correct for intensity, spatial, and other types of dye biases using *robust local regression* (Dudoit and Yang, 2003; Yang et al., 2001, 2002b).

Table 3.2 *Impact of standardization of observations and variables on distance function.*

Distance between observations	Standardized variables	Standardized observations
Euclidean, $w_g = 1$	Changed	Changed
Euclidean, $w_g = 1/s_g^2$	Unchanged	Changed
Mahalanobis	Changed, unless S diagonal	Changed
One-minus-Pearson -correlation	Changed	Unchanged

Note: Recall that by standardizing variables and observations we mean, $x^*_{gi} \leftarrow (x_{gi} - \bar{x}_{g.})/s_g$ and $x^*_{gi} \leftarrow (x_{gi} - \bar{x}_{.i})/s_i$, respectively. Note the relationship between the Euclidean metric $d_E(\cdot, \cdot)$ and the one-minus-Pearson-correlation distance between standardized observations: $d_E(\mathbf{x}^*_i, \mathbf{x}^*_j) = \sqrt{2G(1 - r_{ij})}$, where r_{ij} denotes the correlation between observations i and j.

Distance functions

In classification, observations are assigned to classes on the basis of their *distance* from (or *similarity* to) objects known to be in the classes. The choice of a distance function is most explicit in classifiers such as nearest neighbors. Examples of distance functions are given in Table 3.1. It is often possible to incorporate standardization of the variables as part of the distance function (e.g., general Euclidean distance or Mahalanobis distance). Table 3.2 summarizes the effects of variable (gene) and observation (array) standardization on distances of the L^2 type (i.e., Euclidean and Mahalanobis distances).

Feature selection using the distance function

As described in Section 3.2.3 and Section 3.2.5, appropriate modifications of the distance function, which downweight irrelevant variables based on, for instance, t-statistics, may be used for automatic feature selection.

Impact of standardization and distance function on classification

We noted previously that both the distance function and the scale the observations are measured in can have a large impact on the performance of the classifier. This effect varies depending on the classifier:

Linear and quadratic discriminant analysis. These classifiers are based on the Mahalanobis distance of the observations from the class means. Thus, the classifiers are invariant to standardization of the variables (genes), but not the observations (normalization in the microarray context).

Nearest neighbor classifiers. One must explicitly decide on an appropriate standardization and distance function for the problem under consideration. These classifiers are in general affected by standardization of both features and observations. For example, when using the Euclidean distance between mRNA samples, it might be desirable to scale gene expression measures by their standard deviations (i.e., standardize genes) to prevent genes with large variances from dominating in the classification.

Classification trees. Trees are invariant under monotone transformations of individual features (e.g., gene standardizations introduced previously). They are not invariant to standardization of the observations.

Support vector machines (SVMs). SVMs are based on the Euclidean distance between individual observations (possibly transformed as in the nonlinear case) and a separating hyperplane (margin), they are thus affected by standardization of both features and observations.

3.3.3 Loss function

In many diagnosis settings, the loss incurred from misclassifying a diseased (**d**) person as healthy (**h**) far outweighs the loss incurred by making the error of classifying a healthy person as diseased. These differential misclassification costs should be reflected in the loss function. Suppose the loss from the first error is $e > 1$ times higher than that from the second. For classifiers which can be viewed as estimating the Bayes or maximum likelihood discriminant rules, one could modify posterior probability or likelihood cutoffs accordingly. For example, in the case of the Bayes rule, one could classify a patient as diseased if $p(\mathbf{d}|\mathbf{x}) > c = 1/(1+e)$. Such a classifier would minimize the risk for the loss function $L(\mathbf{d}, \mathbf{h}) = eL(\mathbf{h}, \mathbf{d})$.

The Bayes rule for general loss functions is given in Equation (3.2) in Section 3.2.1. A common situation is when $L(h, l) = L_h I(h \neq l)$, that is, the loss incurred from misclassifying a class h observation is the same irrespective of the predicted class l. Then, the Bayes rule has the form $\mathcal{C}_B(\mathbf{x}) = \text{argmax}_k\, L_k\, p(k|\mathbf{x})$. Any estimator of the class posterior probabilities could be substituted in this equation, and the weights L_k could also be used for ML classifiers, which approximate the Bayes rule for equal priors. The preceding arguments provide a motivation for using weighted votes in nearest neighbor classifiers to reflect differential costs (see Section 3.2.5). A discussion of loss functions for classification trees is given in Section 3.2.6. Some SVM implementations allow specification of class weights to deal with asymmetric class sizes or loss functions (e.g., `class.weight` argument in the `svm` function of the `e1071` R package).

3.3.4 Class representation

Although this may not be immediately obvious, the issues of differential misclassification costs and class priors are closely related (see also Section 3.2.6). In particular, dealing with both issues amounts to imposing different weights or cutoffs for the class posterior probabilities in the Bayes rule or approximations thereof.

Unequal sample class frequencies

In many situations, such as medical diagnosis, the representation of the classes in the learning set does not reflect their importance in the problem. For example, in a binary classification problem with a rare disease class (**d**) and a common healthy class (**h**), a learning set obtained by random sampling from the population would contain a vast majority of healthy cases. Unequal class sample sizes could possibly lead to serious biases in the estimation of posterior probabilities $p(\mathbf{d}|\mathbf{x})$ and $p(\mathbf{h}|\mathbf{x})$. Consider the case of linear discriminant analysis which assumes a common covariance matrix estimated using both samples (see Section 3.2.3). The pooled estimator of variance will be dominated by the more abundant sample. This is fine if the covariance matrix is really the same for both classes. However, in the case of unequal covariance matrices, the bias in the class posterior probabilities $p(k|\mathbf{x})$ is more severe when the classes are unequally represented in the learning set.

The following approaches might help to alleviate estimation biases arising from unequal class representation in the learning set. Bias could be reduced by subsampling the abundant population, so that both classes are on an equal footing in the parameter estimation; however, this would be wasteful of training data when the biased learning set is obtained from a larger learning set. A better approach might be to downweight cases from the abundant class so that the sum of the weights is equal to the number of cases in the less abundant class.

Biased sampling of the classes

Downweighting and subsampling are helpful in dealing with the estimation bias for the class conditional densities $p_k(\mathbf{x})$; however, these approaches effectively make the sample proportions differ from the population proportions and can in turn lead to biased estimators of the class posterior probabilities $p(k|\mathbf{x})$. To see this, let n_k denote the number of class k cases in the learning set. The plug-in estimators of class posterior probabilities are in fact estimating quantities proportional to $p(k|\mathbf{x})n_k/\pi_k$. Adjustment is thus required to ensure that estimators of the class posterior probabilities $p(k|\mathbf{x})$ are approximately unbiased. This can be done by specifying appropriate priors for DA and CART, and by using weighted voting for nearest neighbors.

3.3.5 *Imputation of missing data*

Some classifiers are able to handle missing values by performing automatic imputation (e.g., trees), while others either ignore missing data or require imputed data (e.g., LDA, nearest neighbors).

Classification trees can readily handle missing values through the use of *surrogate splits,* i.e., splits that are most similar to the best split at a particular node (see Section 3.2.6). This is very useful in the analysis of microarray data, where features tend to be highly correlated and a significant proportion of observations are missing. Two goals are accomplished simultaneously by surrogate splits: (i) the tree can be constructed while making use of all the variables, including the ones with missing values; and (ii) a new case can be classified even though some variables are missing. In (i), the goodness of split for a given variable is computed using only observations with no missing values for that variable. In (ii), if the primary split at a node is not defined for a new case, then one looks for the best surrogate variable and uses a splitting rule based on this variable.

For procedures requiring complete data, imputation of missing values must be performed prior to building the classifier. For a detailed study of imputation methods in microarray experiments, the reader is referred to the recent work of Troyanskaya et al. (2001), which suggests that a simple weighted nearest neighbor procedure provides accurate and robust estimates of missing values. In this approach, the neighbors are the genes and the distance between neighbors could be based on the Pearson correlation or Euclidean distance between their gene expression measures across arrays. With k-nearest neighbor imputation, for each gene with missing data: (i) compute its distance to the other $G-1$ genes; (ii) for each missing array, identify the k nearest genes having data for this array and impute the missing entry by the weighted average of the corresponding entries for the k neighbors (here, genes are weighted inversely to their distance from the gene with missing entries). Imputation results were found to be stable and accurate for $k = 10 - 20$ neighbors. Software for k-nearest neighbor imputation is available at `smi-web.stanford.edu/projects/helix/pubs/impute` and in the R package `EMV`.

3.3.6 Polychotomous classification

A number of approaches have been suggested for converting a polychotomous or K-class classification problem into a series of binary or two-class problems. Consideration of binary problems may be advantageous in the case of a large number of classes with unequal representation in the learning set (e.g., lymphoma and NCI 60 datasets in Dettling and Bülmann, 2002).

All $\binom{K}{2}$ pairwise binary classification problems

Friedman (1996a) casts the K-class problem into a series of $\binom{K}{2}$ two-class problems, corresponding to all pairwise class comparisons. In each problem, a separate classification rule is obtained to discriminate between the two classes and to assign an unlabeled test observation to one of the two classes. The final K-class decision rule is obtained by *voting*: the individual class that is selected most often as the predicted class, or *winner*, in the $\binom{K}{2}$ decisions is taken to be the predicted class of the test case. A motivation for such a rule is provided by the simple identity

$\text{argmax}_{1 \le k \le K}\, a_k = \text{argmax}_{1 \le k \le K} \sum_{l=1}^{K} I(a_k > a_l)$, for any set of distinct real numbers $a_1, \ldots, a_K$. Friedman (1996a) shows that the Bayes rule may be reexpressed as

$$\mathcal{C}_B(\mathbf{x}) = \text{argmax}_{1 \le k \le K} \sum_{l=1}^{K} I\left(\frac{p(k \mid \mathbf{x})}{p(k \mid \mathbf{x}) + p(l \mid \mathbf{x})} > \frac{p(l \mid \mathbf{x})}{p(k \mid \mathbf{x}) + p(l \mid \mathbf{x})} \right).$$

Each term in this sum represents an optimal Bayes two-class decision rule. The maximization selects the class with the most winning two-class decisions as the overall prediction at $\mathbf{x}$. Let $f_k^{(kl)} = p(Y = k|\mathbf{x}, Y \in \{k, l\}) = p(k \mid \mathbf{x})/(p(k \mid \mathbf{x}) + p(l \mid \mathbf{x})) = 1 - f_l^{(kl)}$. Then, the previous argument suggests the following general procedure:

$$\mathcal{C}(\mathbf{x}; \mathcal{L}) = \text{argmax}_{1 \le k \le K} \sum_{l=1}^{K} I(\hat{f}_k^{(kl)} > \hat{f}_l^{(kl)}),$$

in which each individual two-class rule is estimated as accurately as possible using $\hat{f}_k^{(kl)}$, for example, by applying model selection techniques. As demonstrated by Friedman (1996a), decomposition of the K-class problem into a series of binary classification rules can lead to substantial gains in accuracy over a K-class rule. Hastie and Tibshirani (1996a) suggest a modification of Friedman's (1996a) rule, pairwise coupling, in which the pairwise class probability estimates are combined into a joint probability estimate for all K classes.

K one-against-all binary classification problems

In a recent article, Dettling and Bülmann (2002) reduce the K-class problem to a series of K binary classification problems of one class against all other classes. Such *one-against-all* approaches are very popular in the machine learning community, where many classifiers are designed only for binary problems (e.g., SVMs). Application of K binary rules to a test case $\mathbf{x}$ yields estimates $\hat{f}_k^{(k)}$ of class posterior probabilities $f_k^{(k)} = p(Y = k|\mathbf{x})$, which may be normalized for the K-class problem: $\hat{p}(k|\mathbf{x}) = \hat{f}_k^{(k)} / \sum_l \hat{f}_l^{(l)}$. The final K-class decision rule selects the class with largest estimated posterior probability $\hat{f}_k^{(k)}$.

3.4 Performance assessment

Different classifiers clearly have different accuracies (i.e., different misclassification rates). In certain medical applications, errors in classification could have serious consequences. For example, when using gene expression data to classify tumor samples, errors could correspond to misdiagnosis and assignment to improper treatment protocol. In this context it is thus essential to obtain reliable estimates of the classification error $p(\mathcal{C}(\mathbf{X}) \neq Y)$ or of other measures of performance. More generally, risk estimation is needed for two main purposes: (i) training the classifier, e.g., selecting an appropriate subset of predictor variables; and (ii) estimating the generalization error of the classifier, i.e., the error rate for a new sample. Different approaches are reviewed next. For a more detailed discussion of performance assessment and of the bias and variance properties of classifiers, the reader is referred to Section 2.7 in Ripley (1996), Friedman (1996b), Breiman (1998a), and Chapter 7 in Hastie et al. (2001).

3.4.1 Bias, variance, and error rates

The random nature of the learning set $\mathcal{L}$ implies that for any realization $\mathbf{x}$ of the feature vector, the predicted class $\mathcal{C}(\mathbf{x};\mathcal{L})$ is a *random variable*. Intuitively, for a fixed value of the feature vector $\mathbf{x}$, as the learning set varies, so will the predicted class $\mathcal{C}(\mathbf{x};\mathcal{L})$. It is thus meaningful and instructive to consider distributional properties (e.g., bias and variance) of classifiers when assessing and comparing their performances. For simplicity, consider binary classification (i.e., $K = 2$ and $Y \in \{-1, 1\}$), and let $f(\mathbf{x}) = p(1|\mathbf{x}) = p(Y = 1|\mathbf{X} = \mathbf{x})$ denote the posterior probability for class 1. Consider predicted classes $\mathcal{C}(\mathbf{x};\mathcal{L})$ based on estimators $\hat{f}(\mathbf{x};\mathcal{L})$ of the class posterior probabilities $f(\mathbf{x})$. Denote the mean and variance of the estimators $\hat{f}(\mathbf{x};\mathcal{L})$ by

$$
\begin{aligned}
\mu_{\mathbf{x}} &= E[\hat{f}(\mathbf{x};\mathcal{L})] \\
\sigma^2_{\mathbf{x}} &= E[(\hat{f}(\mathbf{x};\mathcal{L}) - \mu_{\mathbf{x}})^2].
\end{aligned}
$$

The *mean squared estimation error*, averaged over learning sets $\mathcal{L}$, is

$$
MSE_{\mathbf{x}} = E[(\hat{f}(\mathbf{x};\mathcal{L}) - f(\mathbf{x}))^2] = \sigma^2_{\mathbf{x}} + (\mu_{\mathbf{x}} - f(\mathbf{x}))^2 = \text{Variance} + \text{Bias}^2. \quad (3.6)
$$

Recall that here $\mathbf{x}$ is fixed and the randomness comes from the learning set $\mathcal{L}$. Now consider the effect of $\mu_{\mathbf{x}}$ and $\sigma^2_{\mathbf{x}}$ on classification error (Friedman, 1996b). The Bayes rule is fixed and independent of $\mathcal{L}$, and for binary classification problems it is given by

$$
\mathcal{C}_B(\mathbf{x}) = \begin{cases} 1, & \text{if } f(\mathbf{x}) \geq 1/2 \\ -1, & \text{if } f(\mathbf{x}) < 1/2. \end{cases}
$$

Thus, the classification error for $\mathbf{X} = \mathbf{x}$ is

$$
\begin{aligned}
p(\mathcal{C}(\mathbf{x};\mathcal{L}) \neq Y|\mathbf{X}=\mathbf{x}) &= p(\mathcal{C}(\mathbf{x};\mathcal{L}) = -1)p(Y=1|\mathbf{X}=\mathbf{x}) + p(\mathcal{C}(\mathbf{x};\mathcal{L}) = 1) \\
&\quad\; p(Y=-1|\mathbf{X}=\mathbf{x}) \\
&= p(\mathcal{C}(\mathbf{x};\mathcal{L}) = -1)f(\mathbf{x}) + p(\mathcal{C}(\mathbf{x};\mathcal{L}) = 1)(1 - f(\mathbf{x})) \\
&= |2f(\mathbf{x}) - 1|p(\mathcal{C}(\mathbf{x};\mathcal{L}) \neq \mathcal{C}_B(\mathbf{x})) + p(Y \neq \mathcal{C}_B(\mathbf{x})|\mathbf{X}=\mathbf{x}) \\
&= \text{Estimation error} + \text{Bayes error rate}.
\end{aligned}
$$

Following Friedman (1996b), use a normal approximation for the distribution of $\hat{f}$.

Then,

$$p(\mathcal{C}(\mathbf{x};\mathcal{L}) \neq \mathcal{C}_B(\mathbf{x})) = \Phi\left(\text{sign}(f(\mathbf{x}) - 1/2)\,\frac{1/2 - \mu_{\mathbf{x}}}{\sigma_{\mathbf{x}}}\right),$$

where $\Phi(\cdot)$ is the standard Gaussian cumulative distribution function.

Hence, the classification error rate is

$$p(\mathcal{C}(\mathbf{x};\mathcal{L}) \neq Y|\mathbf{X} = \mathbf{x}) = |2f(\mathbf{x}) - 1|\,\Phi\left(\text{sign}(f(\mathbf{x}) - 1/2)\,\frac{1/2 - \mu_{\mathbf{x}}}{\sigma_{\mathbf{x}}}\right) + p(Y \neq \mathcal{C}_B(\mathbf{x})|\mathbf{X} = \mathbf{x}). \quad (3.7)$$

Comparison of Equation (3.6) and Equation (3.7) shows that moments $\mu_{\mathbf{x}}$ and $\sigma^2_{\mathbf{x}}$ of the distribution of $\hat{f}(\mathbf{x};\mathcal{L})$ have a very different impact on *estimation error* (for $f(\mathbf{x})$) and on *classification error*. In other words, the *bias-variance tradeoff* is very different for these two types of error: for estimation error, the dependence on bias and variance is additive; for classification error, a strong interaction effect occurs. In particular, variance tends to dominate bias for classification error. This suggests that certain methods that have high bias for function estimation may nonetheless perform well for classification because of their low variance. The bias-variance argument provides an explanation for the competitiveness of naive Bayes methods (i.e., DLDA) and nearest neighbor classifiers (Friedman, 1996b).

3.4.2 Resubstitution estimation

In this naive approach, known as *resubstitution error rate estimation* or *training error rate estimation*, the same dataset is used to build the classifier and to assess its performance. That is, the classifier is trained using the entire learning set $\mathcal{L}$, and an estimate of the classification error is obtained by running the *same* learning set $\mathcal{L}$ through the classifier and recording the number of observations with discordant predicted and actual class labels. Although this is a simple approach, the resubstitution error rate can be severely biased downward. Consider the trivial and extreme case when the feature space is partitioned into n sets, each containing a single observation. In this extreme over-fitting situation, the resubstitution error rate is zero; however such a classifier is unlikely to generalize well. That is, the classification error rate as estimated from an independent test set is likely to be high. In general, as the complexity of the classifier increases (i.e., the number of training cycles or epochs increases), the resubstitution error decreases. In contrast, the true generalization error initially decreases but subsequently increases due to over-fitting.

3.4.3 Monte Carlo cross-validation

Suppose a *test set* of labeled observations sampled independently from the same population as the learning set is available. In such a case, an unbiased estimator of the classification error rate may be obtained by running the test set observations through the classifier built from the learning set and recording the proportion of test cases with discordant predicted and actual class labels.

In the absence of a genuine test set, cases in the learning set $\mathcal{L}$ may be divided into two sets, a *training set* $\mathcal{L}_1$ and a *validation set* $\mathcal{L}_2$. The classifier is built using $\mathcal{L}_1$, and the error rate is computed for $\mathcal{L}_2$. It is important to ensure that observations in $\mathcal{L}_1$ and $\mathcal{L}_2$ can be viewed as i.i.d. samples from the population of interest. This can be achieved in practice by randomly dividing the original learning set into two subsets. In addition, to reduce variability in the estimated error rates, this procedure may be repeated a number of times (e.g., 50) and error rates averaged (Breiman, 1998a). A general limitation of this approach is that it reduces effective sample size for training purposes. This is a problem for microarray datasets that have a limited number of observations. No widely accepted guidelines are available for choosing the relative size of these artificial training and validation sets. A possible choice is to leave out a randomly selected 10% of the observations to use as a validation set; however, for comparing the error rates of different classifiers, validation sets containing only 10% of the data are often not large enough to provide adequate discrimination. Increasing validation set size to one-third of the data provides better discrimination in the microarray context.

3.4.4 *Fold cross-validation*

In *V-fold cross-validation* (CV), cases in the learning set $\mathcal{L}$ are randomly divided into V sets $\mathcal{L}_v$, $v = 1, \ldots, V$, of as nearly equal size as possible. Classifiers are built on *training sets* $\mathcal{L} - \mathcal{L}_v$, error rates are computed for the *validation sets* $\mathcal{L}_v$, and averaged over v. A *bias-variance tradeoff* occurs in the selection of V: small Vs typically give a larger bias but a smaller variance.

A commonly used form of CV is *leave-one-out cross-validation* (LOOCV), where $V = n$. LOOCV often results in low bias but high variance estimators of classification error; however, for stable (low variance) classifiers such as k-NN, LOOCV provides good estimators of generalization error (Breiman, 1996b). For large learning sets, LOOCV carries a high computational burden because it requires n applications of the training procedure.

3.4.5 *Bootstrap estimation*

Leave-one-out bootstrap

A number of bootstrap procedures for estimating classification error are reviewed in Ambroise and McLachlan (2002). In the *leave-one-out bootstrap* $B1$, the error rate for learning set case $\mathbf{x}_i$ is obtained from bootstrap samples that do not contain this observation. Bootstrap estimators are typically less variable than LOOCV estimators, and, in particular, the $B1$ estimator can be viewed as a smoothed version of the LOOCV estimator. Because $B1$ is based on a fraction of the original learning set, however, it is upwardly biased ($\approx .632n$ observations are included in each bootstrap sample). A common correction is implemented in the *.632 estimate*: $B.632 = .368RE + .632B1$, where RE is the downwardly biased resubstitution estimate. More general corrections are given by the *.632+ estimate*: $B.632+ = (1-w)RE + wB1$, where the weights w are based on the magnitude of the difference $B1 - RE$.

Out-of-bag estimation

Please refer to Section 3.5.3.

3.4.6 *Honest performance assessment*

Virtually every application of classification methods to microarray data includes some discussion of classifier performance. In such studies, classification error rates, or related measures, are usually reported to: compare different classifiers; select the value of a classifier parameter (e.g., the number of neighbors k in k-NN or the kernel in SVMs); and support statements such as *"clinical outcome X for cancer Y can be predicted accurately based on gene expression measures."* The use of cross-validation (or any other estimation method) is intended to provide accurate estimates of classification error rates. It is important to note that these estimates relate *only* to the experiment that was cross-validated. A common practice in microarray classification is to conduct feature selection by using all of the learning set and then apply cross-validation only to the classifier building portion of the process. In that case, inference can only be applied to the latter part of the process. However, the important genes are usually unknown and the intended inference includes feature selection. Then, the previous CV estimates are downwardly biased and inference is not warranted.

Feature selection *is* an aspect of classifier training, whether done using filter or wrapper methods, and should thus be taken into account in performance assessment. Error estimation procedures should be applied *externally* to the feature selection process, and not *internally* as is commonly done in the microarray literature. When using for example cross-validation to estimate generalization error, features should be selected not on the entire learning set, but separately for each CV training set $\mathcal{L} - \mathcal{L}_v$.

The preceding discussion applies to any error rate estimation procedure and to other aspects of the classifier training process. Examples of classifier parameters that should be included in cross-validation are: the number of predictor variables G; the number of neighbors k and distance function for k-nearest neighbor classifiers; the shrinkage parameter Δ for the nearest shrunken centroid method (Tibshirani et al., 2002); the kernel K and cost parameter C in SVMs. When CV is used to estimate parameters such as the number of genes or the number of neighbors k in k-NN, two rounds of CV may be needed to estimate the generalization error of the classifier. The importance of *honest* cross-validation is demonstrated in Section 3.7 and also discussed in Ambroise and McLachlan (2002) and West et al. (2001).

The approaches described previously can be extended to reflect differential misclassification costs; in such situations, performance is assessed based on the general definition of risk in Equation (3.1). In the case of unequal representation of the classes, some form of stratified sampling may be needed to ensure balance across important classes in all subsamples. In addition, for complex experimental designs, such as factorial or time-course designs, the resampling mechanisms used for computational inference should reflect the design of the experiment.

Finally, note that in the machine learning literature, a frequently employed alternative to risk-based performance measures is the lift. For a given classifier C, the *lift* of a class k is defined as,

$$lift_k = \frac{P(Y = k|C(x) = k)}{P(Y = k)}.$$

The overall lift is obtained by averaging the lifts of individual classes. In general, the greater the lift, the better the classifier.

3.4.7 Confidence in individual predictions

Estimated class posterior probabilities for classifiers based on the Bayes rule may be used to reflect the confidence in predictions for individual observations. Cases with low class posterior probabilities are in general harder to predict and may require follow-up in the form of new laboratory analyses. When estimated class posterior probabilities are not returned explicitly by the classification procedure, one may use derived quantities such as votes for the winning class for k-NN and aggregated predictors using bagging and boosting (see prediction votes and vote margins in Section 3.5.3).

3.5 Aggregating predictors

Breiman (1996a, 1998a) found that gains in accuracy could be obtained by *aggregating predictors* built from perturbed versions of the learning set. In classification, the multiple versions of the predictor are aggregated by *voting*. Let $\mathcal{C}(\cdot; \mathcal{L}_b)$ denote the classifier built from the bth perturbed learning set $\mathcal{L}_b$ and let w_b denote the weight given to predictions made by this classifier. The predicted class for an observation $\mathbf{x}$ is given by

$$\bar{\mathcal{C}}(\mathbf{x}; \mathcal{L}) = \text{argmax}_k \sum_b w_b \, I(\mathcal{C}(\mathbf{x}; \mathcal{L}_b) = k).$$

The *prediction vote* (PV) for a feature vector $\mathbf{x}$ is defined to be

$$PV(\mathbf{x}) = (\max_k \sum_b w_b I(\mathcal{C}(\mathbf{x}; \mathcal{L}_b) = k))/(\sum_b w_b), \text{ and } PV \in [0, 1].$$

When the perturbed learning sets are given equal weights (i.e., $w_b = 1$), the prediction vote is simply the proportion of votes for the winning class. Prediction votes may be used to assess the strength of predictions for individual observations.

The key to improved accuracy is the possible *instability* of a prediction method, i.e., whether small changes in the learning set result in large changes in the predictor (Breiman, 1998a). Voting is a form of averaging, therefore, unstable procedures tend to benefit the most from aggregation. Classification trees tend to be unstable while, for example, nearest neighbor classifiers or DLDA tend to have high bias but low variance. Thus, mainly trees are aggregated in practice. Generally, when trees are combined to create an aggregated classifier, pruning becomes unnecessary and *maximal exploratory trees* are grown until each terminal node contains observations from only one class (Breiman, 1996a, 1998a). Two main classes of methods for generating perturbed versions of the learning set, bagging and boosting, are described next.

3.5.1 Bagging

Standard bagging

In the simplest form of *bagging — bootstrap aggregating* — the perturbed learning sets are nonparametric bootstrap replicates of the learning set (i.e., n observations are drawn at random with replacement from the learning set). Predictors are built for each perturbed dataset and aggregated by plurality voting ($w_b = 1$). A general problem of the nonparametric bootstrap for small datasets is the discreteness of the sampling space. Two methods, described next, get around this problem by sampling from a parametric distribution and by considering convex combinations of the learning set, respectively.

Parametric bootstrap

Perturbed learning sets may be generated from a parametric distribution, for example, according to a mixture of multivariate Gaussian distributions. The mixing probabilities, the mean vectors, and covariance matrices for the class densities may be estimated from the learning set by the corresponding sample quantities.

Convex pseudo-data

Given a learning set $\mathcal{L} = \{(\mathbf{x}_1, y_1), \ldots, (\mathbf{x}_n, y_n)\}$, Breiman (1998b) suggests creating perturbed learning sets based on *convex pseudo-data* (CPD). Each perturbed learning set $\mathcal{L}_b$ is generated by repeating the following n times:

1. Select two instances $(\mathbf{x}, y)$ and $(\mathbf{x}', y')$ at random from the learning set $\mathcal{L}$.
2. Select at random a number v from the interval $[0, d]$, $0 \leq d \leq 1$, and let $u = 1 - v$.
3. The new instance is $(\mathbf{x}'', y'')$, where $y'' = y$ and $\mathbf{x}'' = u\mathbf{x} + v\mathbf{x}'$.

As in standard bagging, a classifier is built for each perturbed learning set $\mathcal{L}_b$ and classifiers are aggregated by plurality voting ($w_b = 1$). Note that when the parameter d is 0, CPD reduces to standard bagging. The larger d, the greater the amount of smoothing. In practice, d could be chosen by cross-validation.

3.5.2 *Random forests*

The term *random forest* refers to a collection of tree classifiers, where each tree depends on the value of a random vector, i.i.d. for all trees in the forest (Breiman, 1999). In random forests, the following sources of randomness are used to generate new predictors:

Random learning set (bagging). Each tree is formed from a bootstrap sample of the learning set — the random vector consists of the outcomes of n draws at random with replacement from $\{1, \ldots, n\}$

Random features. For a fixed parameter $G_0 << G$ (e.g., $G_0 = \sqrt{G}$), G_0 features are randomly selected at each node and only these are searched through for the best split — the random vector consists of the outcomes of G_0 draws at random without replacement from $\{1, \ldots, G\}$.

A maximal exploratory tree is grown (pure terminal nodes) for each bootstrap learning set, and the forest obtains a classification by plurality voting. Note that the main ideas in random forests are applicable to other types of classifiers than trees.

3.5.3 *Byproducts from bagging*

Out-of-bag estimation of error rate

For each bootstrap sample, about one-third ($(1 - 1/n)^n \approx e^{-1} \approx .368$) of the cases are left out and not used in the construction of the tree. These could be used as test set observations for performance assessment purposes. For the bth bootstrap sample, put the *out-of-bag* cases down the bth tree to get a test set classification. For each observation in the learning set, let the final classification of the forest be the class having the most votes for the bootstrap samples in which that observation was out-of-bag. Compare this classification to the class labels of the learning set to get the out-of-bag estimator of the error rate. An unbiased estimator of the misclassification rate for bagged trees is thus obtained automatically as a byproduct of the bootstrap, without the need for subsequent cross-validation or test set estimation (Breiman, 1996c, 1999).

Note that the out-of-bag estimators described previously do not take into account feature selection or other forms of training done prior to aggregating the classification trees. If feature selection is performed beforehand, the out-of-bag estimators of error rates will in general be downwardly biased.

Case-wise information

Other useful byproducts of the bagging procedure are the prediction votes and vote margins:

Prediction votes. The proportions of votes for each class ($\in [0, 1]$) can be viewed as estimates of the class posterior probabilities. The *prediction vote* for the winning class gives a measure of confidence for the prediction of individual observations. Low prediction votes generally correspond to cases that are hard to predict and may require follow-up.

Vote margins. The *vote margin* for a given observation is defined as the proportion of votes for the true class minus the maximum of the proportions of votes for each of the other classes ($\in [-1, 1]$). The lower the vote margin, the poorer the performance of the classifier. Low vote margins may indicate that the corresponding cases are hard to predict or, in some cases, they may reflect mislabeled learning set cases.

Variable importance statistics

In the context of random forests, *variable importance* is defined in terms of the contribution to predictive accuracy (i.e., predictive power). For each tree, randomly permute the values of the gth variable for the out-of-bag cases, put these new covariates down the tree, and get new classifications for the forest. The importance of the gth variable can be defined in a number of ways:

Importance measure 1: the difference between the out-of-bag error rate for randomly permuted gth variable and the original out-of-bag error rate.

Importance measure 2: the average across all cases of the differences between the margins for the randomly permuted gth variable and for the original data.

Importance measure 3: the number of lowered margins minus the number of raised margins.

Importance measure 4: the sum of all decreases in impurity in the forest due to a given variable, normalized by the number of trees.

3.5.4 Boosting

In *boosting*, first proposed by Freund and Schapire (1997), the data are resampled *adaptively* so that the weights in the resampling are increased for those cases most often misclassified. The aggregation of predictors is done by *weighted voting*. Bagging turns out to be a special case of boosting, when the sampling probabilities are uniform at each step and the perturbed predictors are given equal weight in the voting.

In boosting, three elements need to be specified: (i) the type of classifier; (ii) the resampling probabilities and aggregation weights; and (iii) the number of boosting iterations. Any classifier could be used in principle, but unstable classifiers such as classification trees tend to benefit the most from boosting. Boosting procedures generally do not over-fit and are not sensitive to the precise number of iterations in (iii) as long as this number is fairly large, say $B = 50 - 100$. Two popular boosting algorithms, AdaBoost and LogitBoost, are described next.

AdaBoost

We consider an adaptation of Freund and Schapire's *AdaBoost* algorithm, which is described fully in Breiman (1998a) and referred to as *Arc-fs*.

1. **Initialization.** The resampling probabilities $\{p_1^{(0)}, \ldots, p_n^{(0)}\}$ are initialized to be equal (i.e., $p_i^{(0)} = 1/n$).
2. **AdaBoost iterations.** The bth step, $1 \leq b \leq B$, of the algorithm is as follows:
   (a) Using the current resampling probabilities $\{p_1^{(b-1)}, \ldots, p_n^{(b-1)}\}$, sample with replacement from $\mathcal{L}$ to get a learning set $\mathcal{L}_b$ of size n.
   (b) Build a classifier $C(\cdot; \mathcal{L}_b)$ based on $\mathcal{L}_b$.
   (c) Run the learning set $\mathcal{L}$ through the classifier $C(\cdot; \mathcal{L}_b)$ and let $d_i = 1$ if the ith case is classified incorrectly and $d_i = 0$ otherwise.
   (d) Define
   $$\epsilon_b = \sum_i p_i^{(b-1)} d_i \qquad \text{and} \qquad \beta_b = (1 - \epsilon_b)/\epsilon_b,$$
   and update the resampling probabilities for the $(b+1)$st step by
   $$p_i^{(b)} = \frac{p_i^{(b-1)} \beta_b^{d_i}}{\sum_i p_i^{(b-1)} \beta_b^{d_i}}.$$
   (e) In the event that $\epsilon_b \geq 1/2$ or $\epsilon_b = 0$, the resampling probabilities are reset to be equal.
3. **Final classification.** After B steps, the classifiers $C(\cdot; \mathcal{L}_1), \ldots, C(\cdot; \mathcal{L}_B)$ are aggregated by weighted voting, with $C(\cdot; \mathcal{L}_b)$ having weight $w_b = \log(\beta_b)$.

LogitBoost

The *LogitBoost* algorithm was proposed by Friedman et al. (2000) on theoretical grounds, to explain the good performance of boosting-based algorithms. Consider binary classification, with $Y \in \{-1, 1\}$. At each stage, the aggregated classifier can be viewed as an estimate of half the log-odds ratio, $(1/2)\log(f(\mathbf{x})/(1 - f(\mathbf{x})))$, where $f(\mathbf{x}) = p(1|\mathbf{x})$ denotes the posterior probability for class 1. LogitBoost fits an additive logistic model by stage-wise optimization of the binomial log-likelihood.

1. **Initialization.** The weights are initialized to be equal (i.e., $v_i^{(0)} = 1/n$). For observation $\mathbf{x}_i$, $1 \leq i \leq n$, the initial committee function $F^{(0)}(\mathbf{x}_i)$ is set to 0 and the initial probability estimate for class 1 is set to $f^{(0)}(\mathbf{x}_i) = 1/2$.
2. **LogitBoost iterations.** The bth step, $1 \leq b \leq B$, of the algorithm is as follows:
   (a) Building the classifier.
      i. Compute weights and working responses for each observation $i = 1, \ldots, n$

$$v_i^{(b)} = f^{(b-1)}(\mathbf{x}_i)(1 - f^{(b-1)}(\mathbf{x}_i)),$$

$$z_i^{(b)} = \frac{y_i - f^{(b-1)}(\mathbf{x}_i)}{v_i^{(b)}}.$$

      ii. Compute the class predictions $\hat{y}_i^{(b)}$ by weighted least squares fitting of $z_i^{(b)}$ to $\mathbf{x}_i$. For example, in the case of tree stumps $\mathcal{C}$ (Dettling and Bülmann, 2002)

$$\hat{y}_i^{(b)} = \text{argmin}_{\mathcal{C} \in \{stumps\}} \sum_{i=1}^{n} v_i^{(b)} (z_i^{(b)} - \mathcal{C}(\mathbf{x}_i; \mathcal{L}))^2.$$

   (b) Updating

$$\begin{aligned} F^{(b)}(\mathbf{x}_i) &= F^{(b-1)}(\mathbf{x}_i) + \frac{1}{2}\hat{y}_i^{(b)}, \\ \bar{\mathcal{C}}^{(b)}(\mathbf{x}_i; \mathcal{L}) &= \text{sign}(F^{(b)}(\mathbf{x}_i)), \\ f^{(b)}(\mathbf{x}_i) &= (1 + \exp(-2F^{(b)}(\mathbf{x}_i)))^{-1}. \end{aligned}$$

3. **Final classification.** The predicted class for the composite classifier is $\bar{\mathcal{C}}(\mathbf{x}_i; \mathcal{L}) = \bar{\mathcal{C}}^{(B)}(\mathbf{x}_i; \mathcal{L})$.

The LogitBoost algorithm is in general more robust than AdaBoost in the presence of mislabeled observations and inhomogeneities in the learning sample. Dettling and Bülmann (2002) applied the LogitBoost procedure to gene expression data using tree stumps as the weak classifier (i.e., classification trees with only two terminal nodes).

3.6 Datasets

3.6.1 Breast cancer dataset: ER status and lymph node status

This dataset comes from a study of gene expression in breast tumors (West et al., 2001) and is available at `www.genetics.mc.duke.edu/microarray/Published%20work.htm`. Gene expression levels were measured for $n = 49$ breast tumor mRNA samples using Affymetrix high-density oligonucleotide chips containing 7129 human probe sequences (HuGeneFL chips). Two outcomes were measured for each tumor sample: estrogen receptor status, ER+ (25 samples) vs. ER− (24 samples); and lymph node status, affected node present or node+ (25 samples) vs. affected node absent or node − (24 samples). It is believed that different biological mechanisms are involved in the development of breast cancer depending on the ER status of a patient. Gene expression analysis may help to identify those mechanisms by bringing the differences between the ER positive and negative patients to light. Nodal status is an important prognostic factor; patients with no positive nodes generally have better survival. Hence, uncovering the differences between nodal positive and negative patients, and predicting nodal status based on gene expression profiles, may carry important implications for understanding and predicting cancer survival.

Intensity data are available on the Web site and were preprocessed as in Dettling and Bülmann (2002), p.7: (i) thresholding, with a floor of 100 and ceiling of 16,000; (ii) base-10 logarithmic transformation; and (iii) standardization of arrays to have mean zero and unit variance. The data are then summarized by a 7129×49 matrix $X = (x_{gi})$, where x_{gi} denotes the expression measure for gene g in breast tumor sample i.

3.6.2 Brain cancer dataset: medulloblastoma class and survival status

Pomeroy et al. (2002) used Affymetrix oligonucleotide chips to study gene expression in embryonal tumors of the central nervous system (CNS). The data are available at `www-genome.wi.mit.edu/mpr/CNS`; experimental and analysis procedures are described in detail in the "Supplementary Information" document available on the Web site. The authors were interested in distinguishing between different types of brain tumors and predicting survival for patients with medulloblastoma (MD). Although many types of brain tumors are easily distinguished by neurobiologists using histological properties, tumor class assignment still remains somewhat subjective. In particular, the differences between two classes of medulloblastoma, classic and desmoplastic, are subtle and not fully understood. In addition, no accurate markers exist for determining the prognostic status of a patient. Identification of marker genes for predicting survival using expression data would constitute a major achievement in the field of neurobiology. In Pomeroy et al. (2002), gene expression levels were measured for MD tumor mRNA samples using Affymetrix high-density oligonucleotide chips containing a total of 7129 human probe sequences, including 5920 known human gene sequences and 897 ESTs (HuGeneFL chips). Two of the datasets in Pomeroy et al. (2002) are described next.

Dataset B — MD classic vs. desmoplastic

This dataset is based on $n = 34$ medulloblastoma samples, 9 of which are desmoplastic and 25 classic. The data were pre-processed as described in the Supplement, p.69: (i) thresholding, with a floor of 20 and ceiling of 16,000; (ii) filtering, with exclusion of genes with $\max / \min \leq 3$ or $(\max - \min) \leq 100$, where max and min refer respectively to the maximum and minimum intensities for a particular gene across the 34 mRNA samples; and (iii) base-10 logarithmic transformation. The gene expression data were also standardized so that the observations (arrays) have mean 0 and variance 1 across variables (genes). The expression data are then summarized by a 5893×34 matrix $X = (x_{gi})$, where x_{gi} denotes the expression measure for gene g in MD sample i.

Dataset C — MD survival

This dataset contains $n = 60$ medulloblastoma samples, corresponding to 39 survivors and 21 nonsurvivors. The data were preprocessed as described in the Supplement, p.82: (i) thresholding, with a floor of 100 and ceiling of 16,000; (ii) filtering, with exclusion of genes with $\max / \min \leq 5$ or $(\max - \min) \leq 500$, where $\max$ and $\min$ refer respectively to the maximum and minimum intensities for a particular gene across the 60 mRNA samples; and (iii) base-10 logarithmic transformation. The gene expression data were also standardized so that the observations (arrays) have mean 0 and variance 1 across variables (genes). The expression data are then summarized by a 4459×60 matrix $X = (x_{gi})$, where x_{gi} denotes the expression measure for gene g in MD sample i.

3.7 Results

3.7.1 Study design

We have evaluated the performance of different classifiers for predicting tumor class using gene expression data. Classifiers were compared using DNA microarray data from two studies of gene expression in human tumors: the breast cancer study described in West et al. (2001) and the medulloblastoma study of Pomeroy et al. (2002) (details in Section 3.6). For each dataset and each binary outcome of interest, genes were screened according to a two-sample t-statistic (with pooled variance estimator) or a Wilcoxon rank statistic comparing expression measures in each of the two classes. (The Wilcoxon rank statistic is a t-statistic computed using the ranks of the observations in place of their actual values. For more than two classes, one could use an F-statistic; one would also note that, for $K = 2$, this is simply the square of a two-sample t-statistic with pooled estimator of variance. The non-parametric rank analog of the F-statistic is the Kruskal–Wallis statistic.)

In the main comparison (Table 3.3 to Table 3.6, Figure 3.5), error rates were estimated using leave-one-out cross-validation for the following classifiers:

k-nearest neighbor classifiers (k-NN, $k = 1, 3, 5$, Euclidean distance);

Naive Bayes classifiers (diagonal linear discriminant analysis, DLDA, and diagonal quadratic discriminant analysis, DQDA);

LogitBoost ($B = 50,\ 100$);

Random forests ($G_0 = \sqrt{G}, \sqrt{G}/2$);

Support vector machines (SVM, linear and radial kernel, cost $C = 1$ and 100).

Classifiers were built using the G genes with the largest absolute t-statistics or Wilcoxon statistics, $G = 10, 50, 100, 500, 1000$, and all. Gene screening was done separately for each cross-validation training set of $n-1$ observations, and was thus taken into account when assessing the performance of the classifiers. This main comparison allows us to assess the relative merits of different types of classifiers (e.g., k-NN vs. SVMs) and study the sensitivity of the classifiers to the number of features G and classifier parameters (e.g., k for k-NN). In addition, we considered various byproducts of aggregation for random forests, such as out-of-bag error rates, prediction votes, and variable importance statistics.

Finally, we performed a full (or double) cross-validation study to obtain honest estimates of classification error when the training set is used to select classifier parameters such as the number of features G and the number of neighbors k for k-NN. Recall that any training involving parameters that are not specified *a priori* should be taken into account when assessing the performance of the classifier, i.e., should be cross-validated (see Section 3.4). In this full CV study, we only considered k-NN and naive Bayes (DLDA and DQDA) classifiers. Parameters that were part of the cross-validation study for k-NN include the number of genes $G = 10, 50, 100, 500, 1000$, and the number of neighbors $k = 1, 3, 5$. For naive Bayes classifiers, we considered the number of genes $G = 10, 50, 100, 500, 1000$, and DLDA vs. DQDA (i.e., pooled vs. unpooled variance estimators). For each training (leave-one-out) set, a second round of LOOCV was performed in order to determine the combination of parameters that leads to the smallest classification error on that set. The predictor was built using the cross-validated parameters and applied to the observation that was left out at the first level of the LOOCV. The procedure was repeated for each of the n observations and the resulting misclassification rate was recorded, as well as the distribution of classifier parameters such as number of neighbors, type of variance estimator, and number of variables.

3.7.2 Results

Plots of classifier partitions

To get a sense for the types of partitions produced by different classifiers, we considered the two genes with the largest absolute t-statistics for the brain tumor MD survival dataset and built classifiers using only these two genes. Figure 3.1 to Figure 3.4 display the partitions produced by applying linear discriminant analysis (LDA, R `lda` function), quadratic discriminant analysis (QDA, R `qda` function), k-nearest neighbors (R `knn` function, $k = 1, 3, 5, 11$), and CART (R `rpart` function, tenfold CV) to the entire learning set of $n = 60$ tumor samples. In this simple application, the class boundary for QDA is almost linear and the resulting partition is very similar to that of LDA. For k-NN, the boundaries are as expected very irregular with a small number of neighbors k and become smoother as k increases. The CART partition of the two-dimensional feature space is based on two splits only; boundaries are linear

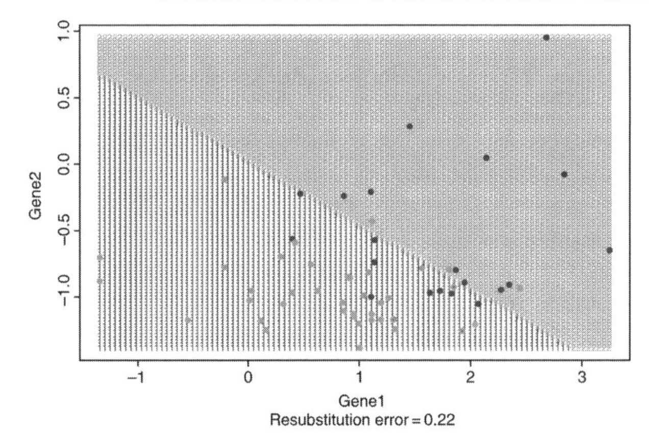

Figure 3.1 *(See color insert following page 114.) Brain tumor MD survival dataset, LDA. Partition produced by linear discriminant analysis (LDA) applied to the two genes with the largest absolute* t*-statistics. Predicted responses "survivor" and "nonsurvivor" are indicated by shades of red and blue, respectively. The entire learning set of* $n = 60$ *samples was used to build the classifier; the resubstitution error rate is shown below the plot. Learning set observations are plotted individually using the color for their true class.*

and parallel to the two axes (splits are based on individual variables in the default CART implementation).

Results of the main comparison and other analyses are described next.

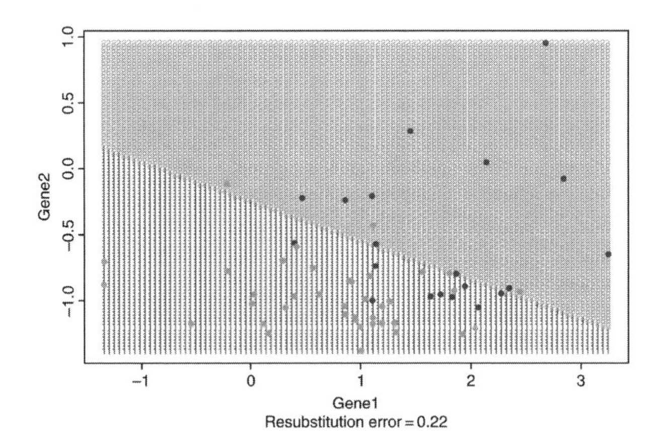

Figure 3.2 *(See color insert following page 114.) Brain tumor MD survival dataset, QDA. Partition produced by quadratic discriminant analysis (QDA) applied to the two genes with the largest absolute* t*-statistics. Predicted responses "survivor" and "nonsurvivor" are indicated by shades of red and blue, respectively. The entire learning set of* $n = 60$ *samples was used to build the classifier; the resubstitution error rate is shown below the plot. Learning set observations are plotted individually using the color for their true class.*

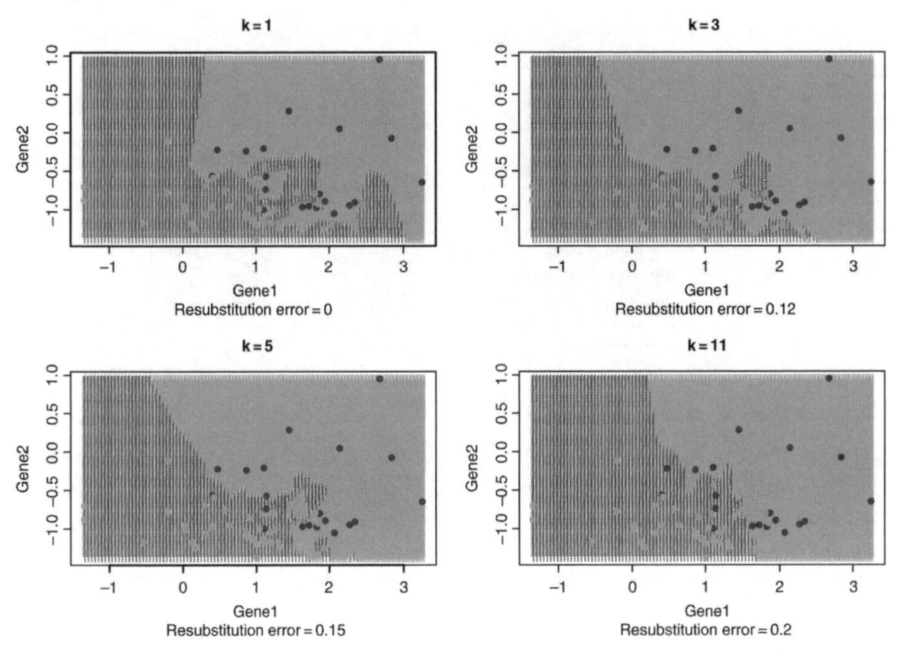

Figure 3.3 *(See color insert following page 114.) Brain tumor MD survival dataset, k-NN. Partitions produced by k-nearest neighbor classification ($k = 1, 3, 5, 11$) applied to the two genes with the largest absolute t-statistics. Predicted responses "survivor" and "nonsurvivor" are indicated by shades of red and blue, respectively. The entire learning set of $n = 60$ samples was used to build the classifier; the resubstitution error rate is shown below the plots. Learning set observations are plotted individually using the color for their true class.*

Good performance of simple classifiers

Diagonal linear discriminant analysis (DLDA), also known as naive Bayes classification, was found to be accurate and robust in the comparison study. A rough minimax argument, in which the best classifier for each dataset is defined to have the smallest maximum error rate, shows that DLDA and k-NN performed very well compared to more sophisticated classifiers (Table 3.3 to Table 3.6, Figure 3.5). The discussion of classification error in terms of bias and variance in Friedman (1996b) provides a more formal explanation for the competitiveness of naive Bayes methods and nearest neighbor classifiers. Although no classifier is uniformly best, it is clearly advantageous to use simple methods, which are intuitive and require little training, unless a more complicated method is demonstrated to be better. As a benchmark, note that the simplest rule would be to predict the class of all observations by the *majority class* (i.e., the class with the largest representation in the learning set). This is related to the Bayes rule when responses Y and feature variables $\mathbf{X}$ are independent (i.e., class posterior probabilities $p(k|\mathbf{x})$ are equal to the priors π_k, and π_k are estimated by the sample proportions). The Bayes risk in this case is $1 - \max_k \pi_k$, and the resubstitution error for the learning set is $1 - \max_k n_k/n$. One can see that for all but the

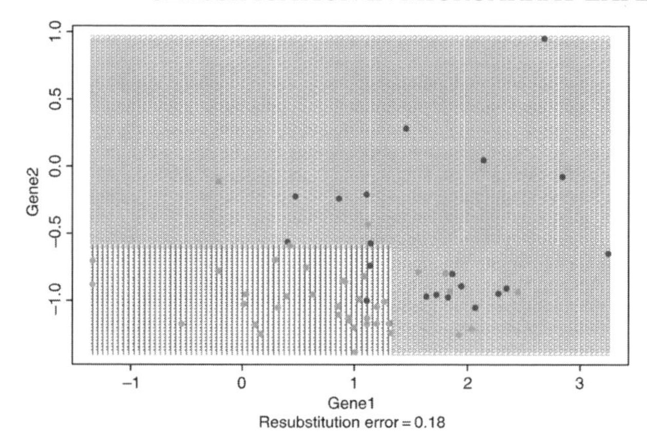

Figure 3.4 *(See color insert following page 114.) Brain tumor MD survival dataset, CART. Partition produced by CART (tenfold CV) applied to the two genes with the largest absolute t-statistics. Predicted responses "survivor" and "nonsurvivor" are indicated by shades of red and blue, respectively. The entire learning set of $n = 60$ samples was used to build the classifier; the resubstitution error rate is shown below the plot. Learning set observations are plotted individually using the color for their true class.*

brain tumor survival dataset, the LOOCV classification error tended to be lower than the benchmark error. In addition, for all but the estrogen dataset, the worst error for any of the classifiers exceeded or approached the benchmark error for the majority rule. Finally, note that random forests had difficulties with unbalanced class sizes and almost always predicted the majority class when unable to build a good predictor. Thus, with the small learning sets currently available in microarray experiments, simple predictors with few tuning parameters are advisable.

Sensitivity to number of features

Feature selection is, in general, a very important issue in classification problems because the choice of features can have a large impact on the performance of the classifiers. Some classifiers, such as trees, perform automatic feature selection and are relatively insensitive to the feature selection scheme. In contrast, standard LDA and nearest neighbor classifiers do not perform feature selection; all variables whether relevant or not are used in building the classifier. Because microarray datasets comprise thousands of features, most of which are likely to be irrelevant for classification purposes, we expected the performance of naive Bayes and k-NN to be very sensitive to the set of genes retained for building the classifier. The results in Table 3.3 to Table 3.6 and Figure 3.5 show only limited changes in performance when varying the number of genes G. The largest effects are observed for the breast tumor nodal dataset, where the smallest number of genes, $G = 10$, yielded the best results for all classifiers. Similar findings are reported in Dettling and Bülmann (2002). To further study the impact of gene screening, we examined the distance matrices used by k-NN

Table 3.3 *Breast tumor estrogen dataset — estimates of classification error using leave-one-out cross-validation.*

Classifier	$G=10$	$G=50$	$G=100$	$G=500$	$G=1000$	$G=7129$ (all)
k-NN						
Euclidean distance						
$k=1$, t-stat	3	6	6	8	**11**	8
$k=1$, W-stat	2	**9**	7	**9**	**9**	8
$k=3$, t-stat	5	5	4	**7**	5	6
$k=3$, W-stat	5	5	4	**6**	5	**6**
$k=5$, t-stat	5	**6**	5	3	4	5
$k=5$, W-stat	4	4	4	4	**5**	**5**
DLDA						
t-stat	**6**	3	4	4	4	**6**
W-stat	**6**	4	4	4	4	**6**
DQDA						
t-stat	5	4	6	6	7	**15**
W-stat	6	4	4	6	7	**15**
LogitBoost						
$B=50$, t-stat	**4**	2	2	3	3	2
$B=50$, W-stat	**4**	2	3	**4**	3	2
$B=100$, t-stat	4	1	2	3	**5**	3
$B=100$, W-stat	**4**	3	2	3	**4**	3
Random forest						
$G_0=\sqrt{G}/2$, t-stat	3	3	**5**	4	4	**5**
$G_0=\sqrt{G}/2$, W-stat	3	4	**5**	**5**	**5**	**5**
$G_0=\sqrt{G}$, t-stat	3	3	**5**	**5**	**5**	4
$G_0=\sqrt{G}$, W-stat	2	3	4	**5**	**5**	4
SVM						
linear, $C=1$, t-stat	5	4	4	**7**	6	6
linear, $C=1$, W-stat	**7**	3	3	**7**	6	6
linear, $C=100$, t-stat	5	4	4	**7**	6	6
linear, $C=100$, W-stat	**7**	3	3	**7**	6	6
radial, $C=1$, t-stat	**6**	5	**6**	5	5	4
radial, $C=1$, W-stat	5	**6**	5	**6**	4	4
radial, $C=100$, t-stat	6	5	5	**7**	6	6
radial, $C=100$, W-stat	5	**6**	3	**6**	**6**	**6**

Note: The G genes with the largest absolute t- and Wilcoxon statistics for comparing expression measures in ER+ and ER− patients were selected separately for each CV training set of size $n-1=48$. The majority rule benchmark resubstitution error is 24. The maximum error for each predictor (row of the table) is enclosed in a box.

Table 3.4 *Breast tumor nodal dataset — estimates of classification error using leave-one-out cross-validation.*

Classifier	$G=10$	$G=50$	$G=100$	$G=500$	$G=1000$	$G=7129$ (all)
k-NN						
Euclidean distance						
$k=1$, t-stat	8	**25**	21	19	19	24
$k=1$, W-stat	10	16	14	21	**24**	**24**
$k=3$, t-stat	8	22	19	19	22	**26**
$k=3$, W-stat	10	16	18	19	21	**26**
$k=5$, t-stat	8	22	18	22	19	**24**
$k=5$, W-stat	10	13	19	19	21	**24**
DLDA						
t-stat	8	14	**19**	17	15	17
W-stat	10	10	15	**17**	16	**17**
DQDA						
t-stat	11	17	18	22	24	**26**
W-stat	12	13	22	22	20	**26**
LogitBoost						
$B=50$, t-stat	9	17	**18**	16	15	13
$B=50$, W-stat	9	12	**19**	12	8	13
$B=100$, t-stat	9	**21**	18	14	14	12
$B=100$, W-stat	9	13	**21**	11	11	12
Random forest						
$G_0=\sqrt{G}/2$, t-stat	6	13	15	18	16	**20**
$G_0=\sqrt{G}/2$, W-stat	9	9	17	16	15	**20**
$G_0=\sqrt{G}$, t-stat	7	13	14	15	17	**18**
$G_0=\sqrt{G}$, W-stat	8	9	15	17	17	**18**
SVM						
linear, $C=1$, t-stat	7	14	**21**	12	15	20
linear, $C=1$, W-stat	9	17	16	14	14	**20**
linear, $C=100$, t-stat	9	14	**21**	12	15	20
linear, $C=100$, W-stat	11	18	16	14	14	**20**
radial, $C=1$, t-stat	8	16	18	15	16	**29**
radial, $C=1$, W-stat	7	12	15	14	17	**29**
radial, $C=100$, t-stat	8	16	19	15	14	**20**
radial, $C=100$, W-stat	9	16	14	14	15	**20**

Note: The G genes with the largest absolute t- and Wilcoxon statistics for comparing expression measures in nodal+ and nodal− patients were selected separately for each CV training set of size $n-1=48$. The majority rule benchmark resubstitution error is 24. The maximum error for each predictor (row of the table) is enclosed in a box.

Table 3.5 *Brain tumor classic vs. desmoplastic MD dataset — estimates of classification error using leave-one-out cross-validation.*

Classifier	$G=10$	$G=50$	$G=100$	$G=500$	$G=1000$	$G=5893$ (all)
k-NN						
Euclidean distance						
$k=1$, t-stat	6	6	5	4	3	**10**
$k=1$, W-stat	4	6	4	4	4	**10**
$k=3$, t-stat	5	5	3	4	5	**8**
$k=3$, W-stat	4	4	3	4	4	**8**
$k=5$, t-stat	6	3	3	3	5	**8**
$k=5$, W-stat	5	4	4	4	6	**8**
DLDA						
t-stat	**6**	4	5	4	4	5
W-stat	4	**5**	**5**	4	4	**5**
DQDA						
t-stat	6	6	6	8	8	**9**
W-stat	5	6	5	5	**9**	**9**
LogitBoost						
$B=50$, t-stat	**12**	10	8	6	6	7
$B=50$, W-stat	5	3	6	8	**7**	**7**
$B=100$, t-stat	**12**	7	8	7	7	9
$B=100$, W-stat	6	5	7	8	**9**	**9**
Random forest						
$G_0=\sqrt{G}/2$, t-stat	5	6	6	8	**9**	**9**
$G_0=\sqrt{G}/2$, W-stat	4	7	7	8	**9**	**9**
$G_0=\sqrt{G}$, t-stat	6	6	6	8	**9**	**9**
$G_0=\sqrt{G}$, W-stat	5	7	7	**9**	**9**	**9**
SVM						
linear, $C=1$, t-stat	5	**8**	7	7	6	6
linear, $C=1$, W-stat	**8**	7	7	6	5	6
linear, $C=100$, t-stat	4	**8**	7	7	6	6
linear, $C=100$, W-stat	**9**	7	7	6	5	6
radial, $C=1$, t-stat	5	5	4	6	6	**9**
radial, $C=1$, W-stat	4	6	4	5	6	**9**
radial, $C=100$, t-stat	6	**7**	**7**	6	5	6
radial, $C=100$, W-stat	5	**7**	5	5	6	6

Note: The G genes with the largest absolute t- and Wilcoxon statistics for comparing expression measures in classic vs. desmoplastic MD patients were selected separately for each CV training set of size $n-1=33$. The majority rule benchmark resubstitution error is 9. The maximum error for each predictor (row of the table) is enclosed in a box.

Table 3.6 *Brain tumor MD survival dataset — estimates of classification error using leave-one-out cross-validation.*

Classifier	$G=10$	$G=50$	$G=100$	$G=500$	$G=1000$	$G=4459$ (all)
k**-NN**						
Euclidean distance						
$k=1$, t-stat	20	29	**32**	25	26	28
$k=1$, W-stat	18	24	24	25	**28**	**28**
$k=3$, t-stat	20	**24**	22	20	19	23
$k=3$, W-stat	16	22	19	18	21	**23**
$k=5$, t-stat	**24**	21	22	22	18	23
$k=5$, W-stat	19	22	19	20	22	**23**
DLDA						
t-stat	19	25	23	24	24	**26**
W-stat	17	**26**	21	24	25	**26**
DQDA						
t-stat	22	24	23	**25**	23	23
W-stat	17	**27**	22	**27**	26	23
LogitBoost						
$B=50$, t-stat	22	21	**24**	21	21	18
$B=50$, W-stat	17	23	**25**	20	17	18
$B=100$, t-stat	19	19	**26**	16	16	16
$B=100$, W-stat	21	24	**26**	19	15	16
Random forest						
$G_0=\sqrt{G}/2$, t-stat	17	**21**	20	19	**21**	**21**
$G_0=\sqrt{G}/2$, W-stat	16	**21**	20	20	20	**21**
$G_0=\sqrt{G}$, t-stat	18	21	20	19	20	**22**
$G_0=\sqrt{G}$, W-stat	17	20	18	19	20	**22**
SVM						
linear, $C=1$, t-stat	17	**29**	**29**	22	22	19
linear, $C=1$, W-stat	16	**29**	25	24	24	19
linear, $C=100$, t-stat	19	**29**	**29**	22	22	19
linear, $C=100$, W-stat	21	**28**	25	24	24	19
radial, $C=1$, t-stat	20	20	20	**22**	17	21
radial, $C=1$, W-stat	14	21	16	**24**	18	21
radial, $C=100$, t-stat	20	**25**	**25**	22	21	20
radial, $C=100$, W-stat	21	**27**	23	24	23	20

Note: The G genes with the largest absolute t- and Wilcoxon statistics for comparing expression measures in MD survivors vs. nonsurvivors were selected separately for each CV training set of size $n-1=59$. The majority rule benchmark resubstitution error is 21. The maximum error for each predictor (row of the table) is enclosed in a box.

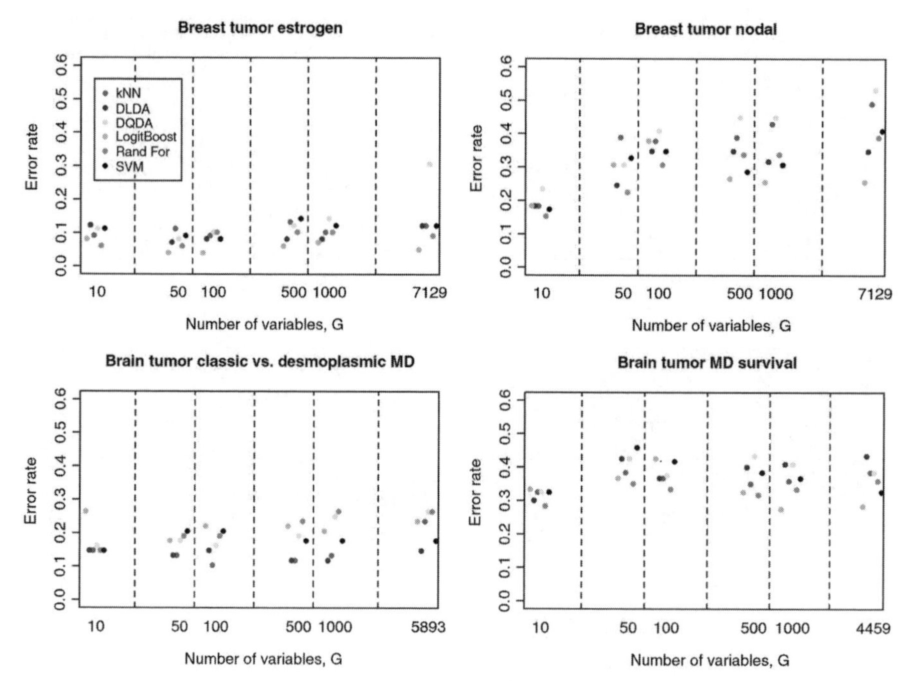

Figure 3.5 *(See color insert following page 114.) Comparison of classification error rates. Classification error rates were estimated for each classifier using leave-one-out cross-validation as in Table 3.3 to Table 3.6. For each family of classifiers and given number of genes G, median error rates were computed over the parameter values considered in Table 3.3 to Table 3.6. The median error rates are plotted vs. the number of genes G (on log-scale) for each family of classifiers.*

(color images in Figure 3.6 and Figure 3.7). Increasing the number of genes G clearly attenuates between class differences, especially for the estrogen dataset; however, the most similar samples are still found within classes. Thus, k-NN classifiers with a small number of neighbors k had a reasonable performance even with the full complement of genes. (We chose to display Pearson correlation matrices and not Euclidean distance matrices because the former are easier to compare across datasets. In addition, for standardized arrays, classification results based on Euclidean distance and correlation matrices are very similar and identical for $G =$ all.) Random forests and LogitBoost were, as expected, not very sensitive to the number of features G. Although classifier performance does not always deteriorate with an increasing number of features, some screening of the genes down to, for example, $G = 10 - 100$, seems advisable.

Feature selection procedure

In the main comparison, genes were screened using a t- or Wilcoxon statistic and the G genes with the largest absolute test statistics were retained to build the classifier.

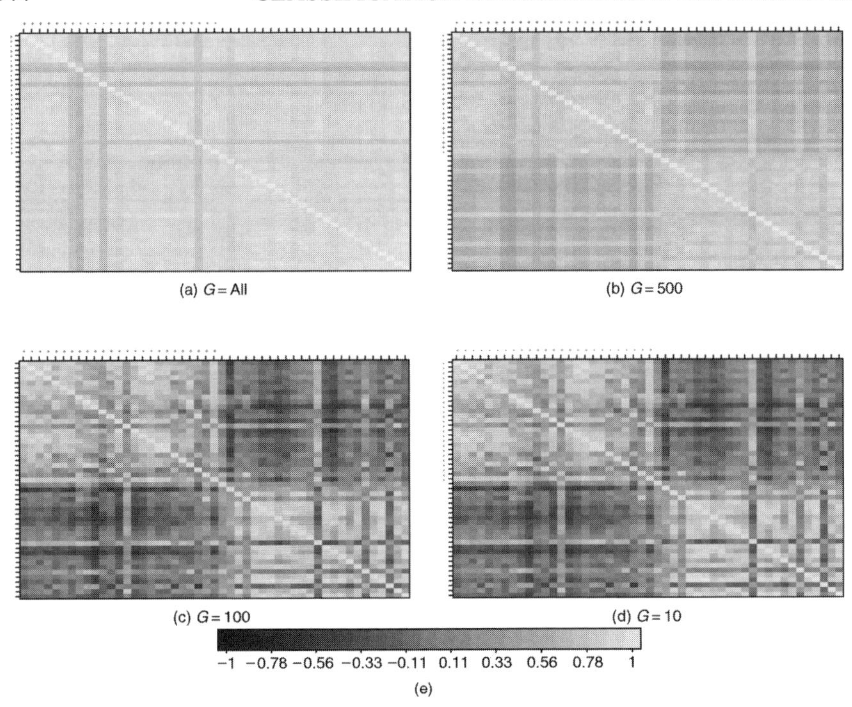

Figure 3.6 *(See color insert following page 114.) Breast tumor estrogen dataset. Images of the correlation matrix for the 49 breast tumor mRNA samples based on expression profiles for: (a) all* $G = 7129$ *genes; (b) the* $G = 500$ *genes with the largest absolute t-statistics; (c) the* $G = 100$ *genes with the largest absolute t-statistics; (d) the* $G = 10$ *genes with the largest absolute t-statistics. The mRNA samples are ordered by class: first ER positive, then ER negative. Increasingly positive correlations are represented with yellows of increasing intensity, and increasingly negative correlations are represented with blues of increasing intensity. The color bar below the images may be used for calibration purposes.*

In general, gene screening based on robust Wilcoxon statistics did not seem to consistently increase or decrease error rates compared to using t-statistics. The largest effects were seen for the breast nodal dataset. Naive one-gene-at-a-time approaches do not account for the joint effects of genes on class distinction and can produce groups of highly correlated genes providing redundant information on classification. In contrast, variable importance statistics derived from random forests account for the joint action of genes and could be used in principle to detect genes with weak main effects but strong interactions.

We thus compared t-statistics and Wilcoxon statistics to two of the importance statistics described in Section 3.5.3. The R function `RanForests`, for Random Forests version 1, only returns measures 1 and 4. Importance measure 1 is very discrete (it refers to numbers of observations) and hardly provides any discrimination between genes. Measure 4 provides better discrimination and qualitatively matches the t- and

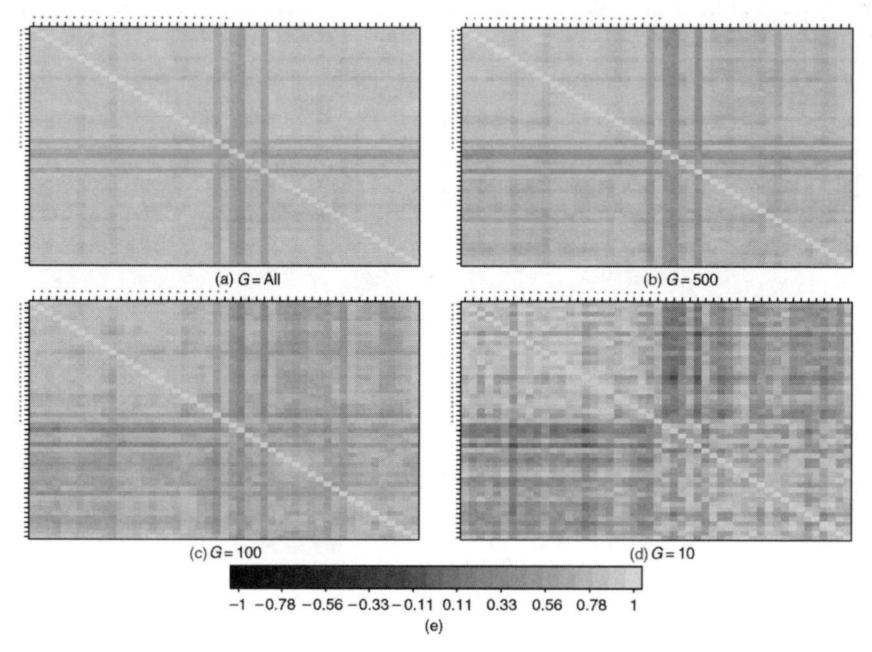

Figure 3.7 *(See color insert following page 114.) Breast tumor nodal dataset. Images of the correlation matrix for the 49 breast tumor mRNA samples based on expression profiles for: (a) all* $G = 7129$ *genes; (b) the* $G = 500$ *genes with the largest absolute t-statistics; (c) the* $G = 100$ *genes with the largest absolute t-statistics; (d) the* $G = 10$ *genes with the largest absolute t-statistics. The mRNA samples are ordered by class: first nodal positive, then nodal negative. Increasingly positive correlations are represented with yellows of increasing intensity, and increasingly negative correlations are represented with blues of increasing intensity. The color bar below the images may be used for calibration purposes.*

Wilcoxon statistics for high values (Figure 3.8 to Figure 3.11). In other applications of random forests, we have found that measure 2 is similar to measure 4, while measure 3 is similar to measure 1.

Sensitivity to classifier parameters

In addition to the number of genes G, most classifiers in this study have a number of tuning parameters with values that affect classification error. In a previous article, we found that nearest neighbor classifiers performed best with a small number of neighbors, $k = 1 - 5$ (Dudoit et al., 2002a). Furthermore, considering a much larger number of neighbors is often infeasible in microarray experiments due to small sample sizes. In the present study, we considered only small k's between 1 and 5, and found that, within this range, classifier performance was fairly robust to the choice of k. For SVMs, no single kernel was found to be best across datasets and number of genes G. The performance of SVMs was fairly insensitive to the cost parameter or penalty C. Random forests were robust to the parameter G_0 for the number of random

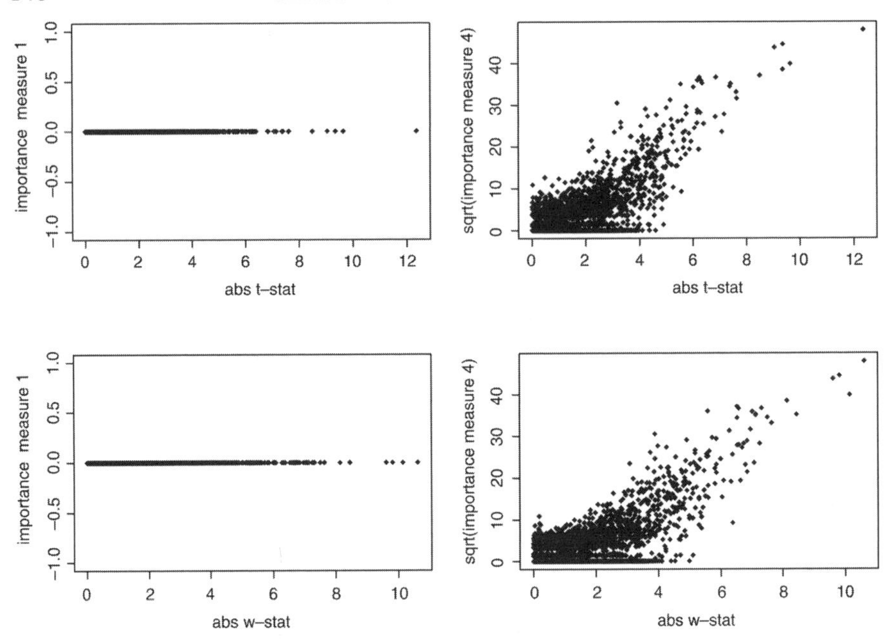

Figure 3.8 *Breast tumor estrogen dataset. Comparison of two-sample* t*-statistics, Wilcoxon statistics, and random forests importance measures 1 and 4 (*$G_0 = \sqrt{G}$*). Test statistics were computed using the entire learning set of* $n = 49$ *observations. Plot of importance measure 1 vs. absolute* t*-statistic (top left); plot of square-root of importance measure 4 vs. absolute* t*-statistic (top right); plot of importance measure 1 vs. absolute Wilcoxon statistic (bottom left); plot of square root of importance measure 4 vs. absolute Wilcoxon statistic (bottom right).*

features considered at each node. The number of boosting iterations B for LogitBoost classifiers did not have a strong effect on classification error; $B = 50$ or 100 appears to be a good choice. This observation is consistent with the results of Dettling and Bülmann (2002), who concluded that the advantages of a more refined procedure for selecting B based on LOOCV were limited. It is very important to note that only gene screening was cross-validated in the main comparison, and not the selection of other parameters, such as the number of neighbors in k-NN. Although the error rates in Table 3.3 to Table 3.6 are meaningful for comparison purposes, they are not good estimates of generalization error, as discussed in Section 3.4 and next. For accurate estimates of generalization error, one would need to make all training decisions on the training set only (i.e., on each set of $n - 1$ training observations for LOOCV).

Impact of aggregation

As expected, bagging improved the performance of unstable classifiers such as trees, but appears to have very little effect on stable classifiers such as k-NN and naive Bayes (data not shown).

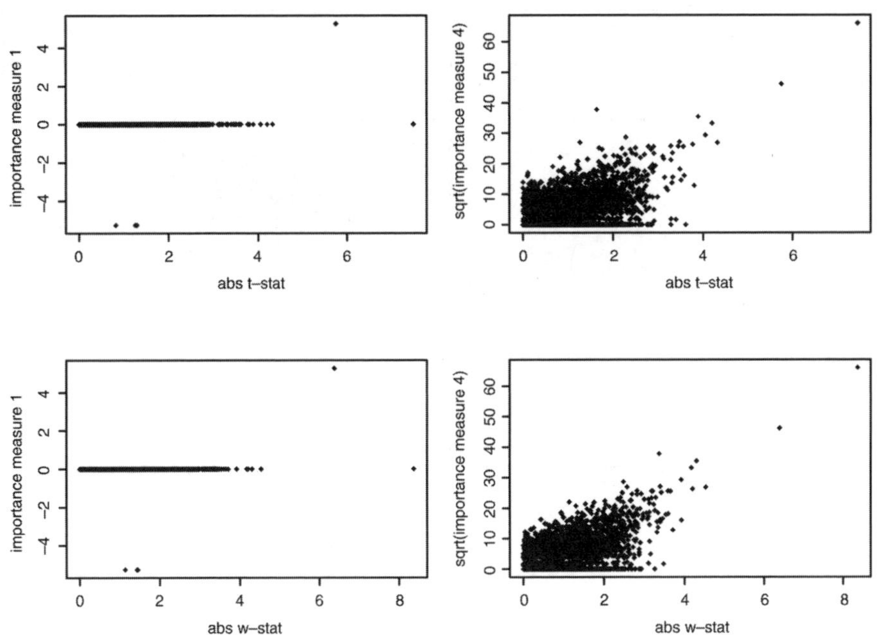

Figure 3.9 *Breast tumor nodal dataset. Comparison of two-sample t-statistics, Wilcoxon statistics, and random forests importance measures 1 and 4 ($G_0 = \sqrt{G}$). Test statistics were computed using the entire learning set of $n = 49$ observations. Plot of importance measure 1 vs. absolute t-statistic (top left); plot of square-root of importance measure 4 vs. absolute t-statistic (top right); plot of importance measure 1 vs. absolute Wilcoxon statistic (bottom left); plot of square root of importance measure 4 vs. absolute Wilcoxon statistic (bottom right).*

Prediction votes

Predictions votes, reflecting the confidence of predictions for individual observations, are returned as byproducts of aggregated classifiers (random forests). The boxplots of prediction votes for correct and incorrect predictions in Figure 3.12 indicate that larger votes tend to be associated with correct predictions. The breast tumor nodal and brain survival datasets tended to have high classification error rates. The low confidence in the predictions for random forests is reflected in low prediction votes for both correct and incorrect classifications (around the minimum value of 0.5 for binary classification). Thus, as suggested in our earlier study (Dudoit et al., 2002a), low prediction votes tend to correspond to cases which are hard to predict and may require follow-up.

Out-of-bag error rate

Another useful byproduct of bagged classifiers is the out-of-bag estimate of classification error. We found this error rate to be very similar to the LOOCV estimate

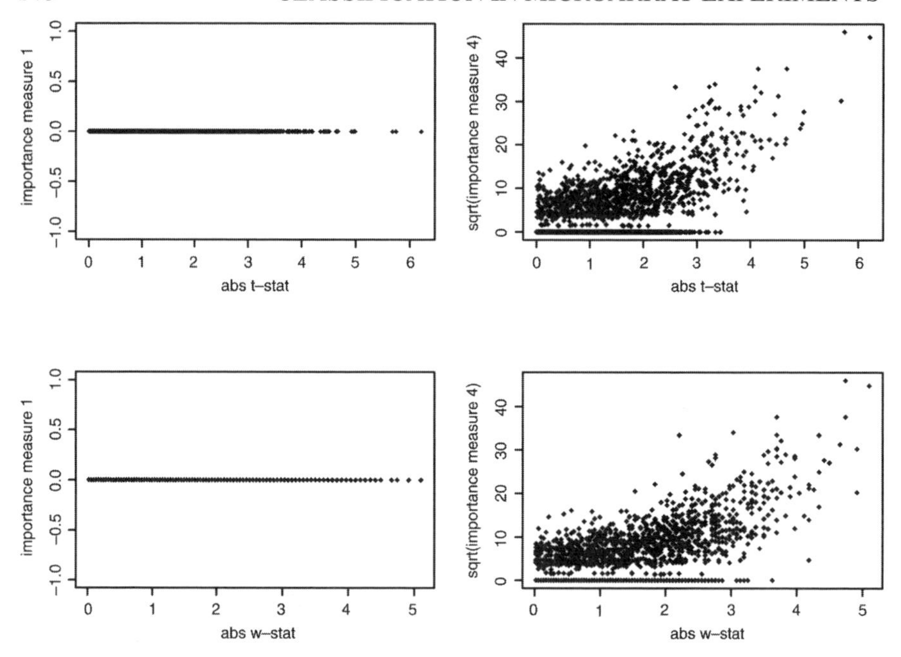

Figure 3.10 *Brain tumor classic vs. desmoplastic MD dataset. Comparison of two-sample t-statistics, Wilcoxon statistics, and random forests importance measures 1 and 4 ($G_0 = \sqrt{G}$). Test statistics were computed using the entire learning set of $n = 34$ observations. Plot of importance measure 1 vs. absolute t-statistic (top left); plot of square-root of importance measure 4 vs. absolute t-statistic (top right); plot of importance measure 1 vs. absolute Wilcoxon statistic (bottom left); plot of square-root of importance measure 4 vs. absolute Wilcoxon statistic (bottom right).*

for classifiers built using the entire set of genes (Table 3.7). Out-of-bag estimation is attractive because it is part of the classifier building process and does not require a separate test set; however, it does not take into account feature selection or other forms of training done prior to aggregating the classifier. If feature selection is performed beforehand, the out-of-bag estimates of error rates will in general be biased downward (Dudoit et al., 2002a). Incorporating some form of feature selection as part of the bagging process could prove useful.

Honest estimates of error rates

Last but not least, *honest* classifier performance assessment is a very important notion that has been ignored to a large extent in microarray data analysis. It is common practice in microarray classification to screen genes and fine-tune classifier parameters (e.g., number of neighbors k in nearest neighbor classification) using *all* of the learning set and then perform cross-validation only on the classifier building portion of the process. The resulting estimates of generalization error are generally biased downward and thus overly optimistic. The importance of taking into account gene screening

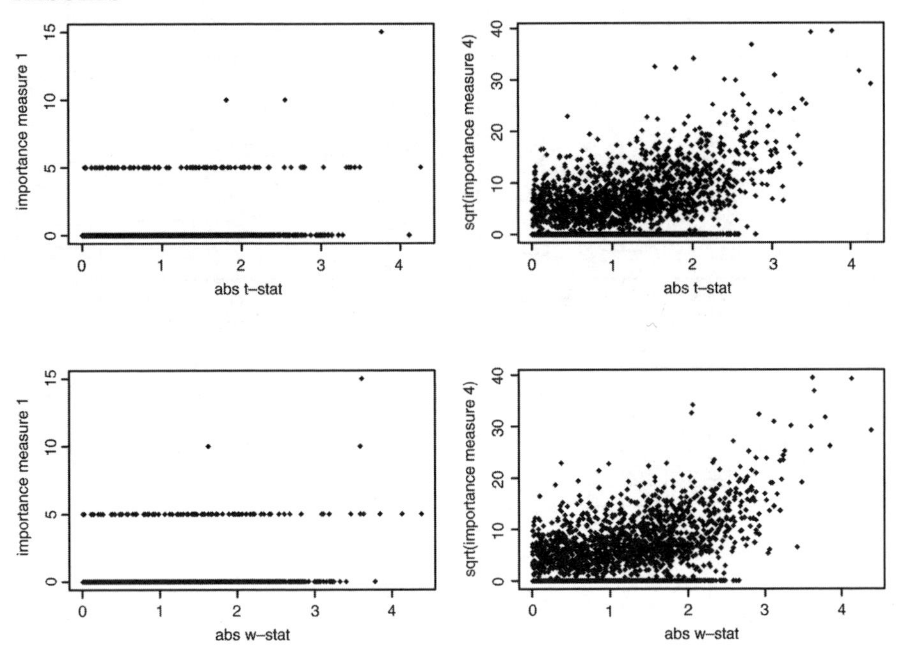

Figure 3.11 *Brain tumor MD survival dataset. Comparison of two-sample t-statistics, Wilcoxon statistics, and random forests importance measures 1 and 4 ($G_0 = \sqrt{G}$). Test statistics were computed using the entire learning set of $n = 60$ observations. Plot of importance measure 1 vs. absolute t-statistic (top left); plot of square-root of importance measure 4 vs. absolute t-statistic (top right); plot of importance measure 1 vs. absolute Wilcoxon statistic (bottom left); plot of square-root of importance measure 4 vs. absolute Wilcoxon statistic (bottom right).*

and other training decisions in error rate estimation procedures was also stressed by Ambroise and McLachlan (2002) and West et al. (2001).

The results in Table 3.8 to Table 3.11 contrast LOOCV estimates of classification error for genes selected using the entire learning set $\mathcal{L}$ and for genes selected separately for each CV training set of size $n - 1$. The later estimates are generally much higher and more realistic estimates of generalization error. The difference is particularly striking for the breast tumor nodal and brain tumor survival datasets. Figure 3.13 compares the classification error estimates of 1-NN and DLDA, for three different estimation procedures applied to the breast tumor nodal dataset:

1. *Resubstitution estimation*, where the entire learning set is used to perform feature selection, build the classifier, and estimate classification error.
2. *Internal LOOCV*, where feature selection is done on the entire learning set and LOOCV is applied only to the classifier building process.
3. *External LOOCV*, where LOOCV is applied to the feature selection *and* the classifier building process.

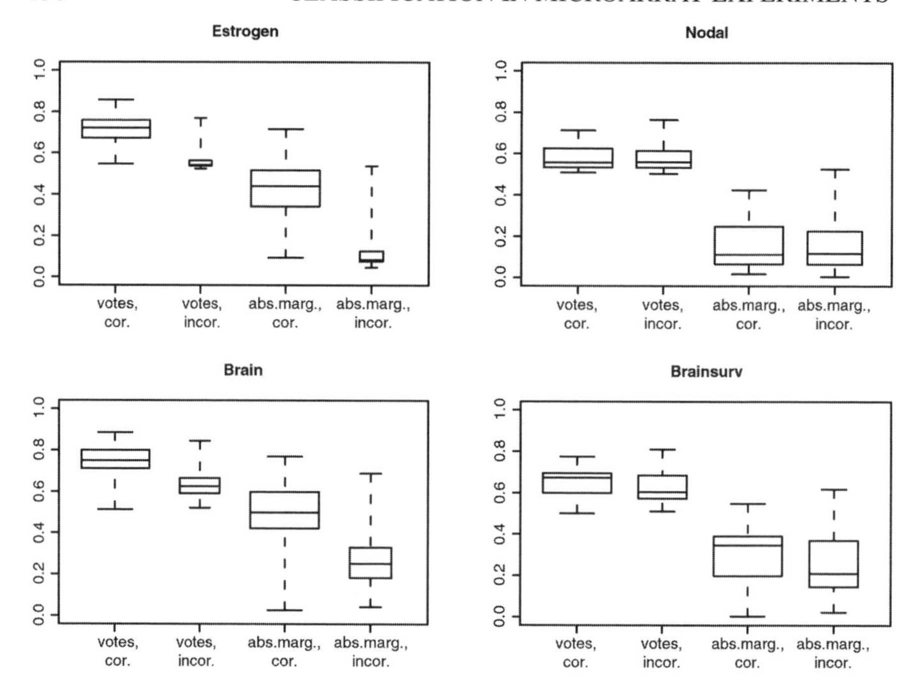

Figure 3.12 *Prediction votes and vote margins. The boxplots summarize prediction votes and absolute vote margins obtained by applying random forests to the entire learning set for each of the four datasets (i.e., all genes and all samples),* $G_0 = \sqrt{G}$. *Prediction votes and margins are stratified according to the correctness of the prediction. The widths of the boxplots are proportional to the square root of the number of observations in each of the groups.*

This analysis again demonstrates the large downward biases resulting from improper cross-validation.

In order to obtain honest estimates of generalization error, one should perform a full cross-validation of gene screening and other parameter training. The results of a full

Table 3.7 *Out-of-bag error rates.* [a]

Dataset	Out-of-bag error rate	LOOCV
Breast: Estrogen	$5/49$	$5/49$
Breast: Nodal	$19/49$	$20/49$
Brain: MD classic vs. desmoplastic	$9/34$	$9/34$
Brain: MD survival	$20/60$	$21/60$

[a] Out-of-bag error rates from random forests built using the entire learning set for each of the four datasets (i.e., all genes and all samples), $G_0 = \sqrt{G}$.

Table 3.8 *Breast tumor estrogen dataset — estimates of classification error using leave-one-out cross-validation, with gene screening by t-statistic performed on the entire learning set*[a] *or separately for each CV training set.*[b]

	$G = 10$	$G = 50$	$G = 100$	$G = 500$	$G = 1000$
k-NN, $k = 1$					
Euclidean distance					
Internal	3	6	6	6	7
External	3	6	6	8	11
DLDA					
Internal	2	1	2	2	3
External	6	3	4	4	4
Random forest					
$G_0 = \sqrt{G}/2$					
Internal	2	2	2	5	3
External	3	3	5	4	4

[a] Internal LOOCV, $n = 49$ cases used for feature selection.

[b] External LOOCV, $n - 1 = 48$ cases used for feature selection.

Table 3.9 *Breast tumor nodal dataset — estimates of classification error using leave-one-out cross-validation, with gene screening by t-statistic performed on the entire learning set*[a] *or separately for each CV training set.*[b]

	$G = 10$	$G = 50$	$G = 100$	$G = 500$	$G = 1000$
k-NN, $k = 1$					
Euclidean distance					
Internal	6	6	12	14	15
External	8	25	21	19	19
DLDA					
Internal	5	6	5	5	6
External	8	14	19	17	15
Random forest					
$G_0 = \sqrt{G}/2$					
Internal	5	6	6	7	10
External	6	13	15	18	16

[a] Internal LOOCV, $n = 49$ cases used for feature selection.

[b] External LOOCV, $n - 1 = 48$ cases used for feature selection.

Table 3.10 *Brain tumor classic vs. desmoplastic MD dataset — estimates of classification error using leave-one-out cross-validation, with gene screening by t-statistic performed on the entire learning set*[a] *or separately for each CV training set.*[b]

	$G = 10$	$G = 50$	$G = 100$	$G = 500$	$G = 1000$
k-**NN,** $k = 1$					
Euclidean distance					
Internal	3	2	2	2	1
External	6	6	5	4	3
DLDA					
Internal	2	2	2	2	3
External	6	4	5	4	4
Random forest					
$G_0 = \sqrt{G}/2$					
Internal	2	2	3	5	5
External	5	6	6	8	9

[a] Internal LOOCV, $n = 34$ cases used for feature selection.

[b] External LOOCV, $n - 1 = 33$ cases used for feature selection.

Table 3.11 *Brain tumor MD survival dataset — estimates of classification error using leave-one-out cross-validation, with gene screening by t-statistic performed on the entire learning set*[a] *or separately for each CV training set.*[b]

	$G = 10$	$G = 50$	$G = 100$	$G = 500$	$G = 1000$
k-**NN,** $k = 1$					
Euclidean distance					
Internal	17	10	16	18	19
External	20	29	32	25	26
DLDA					
Internal	9	7	6	12	14
External	19	25	23	24	24
Random forest					
$G_0 = \sqrt{G}/2$					
Internal	7	11	13	12	16
External	17	21	20	19	21

[a]Internal LOOCV, $n = 60$ cases used for feature selection.

[b]External LOOCV, $n - 1 = 59$ cases used for feature selection.

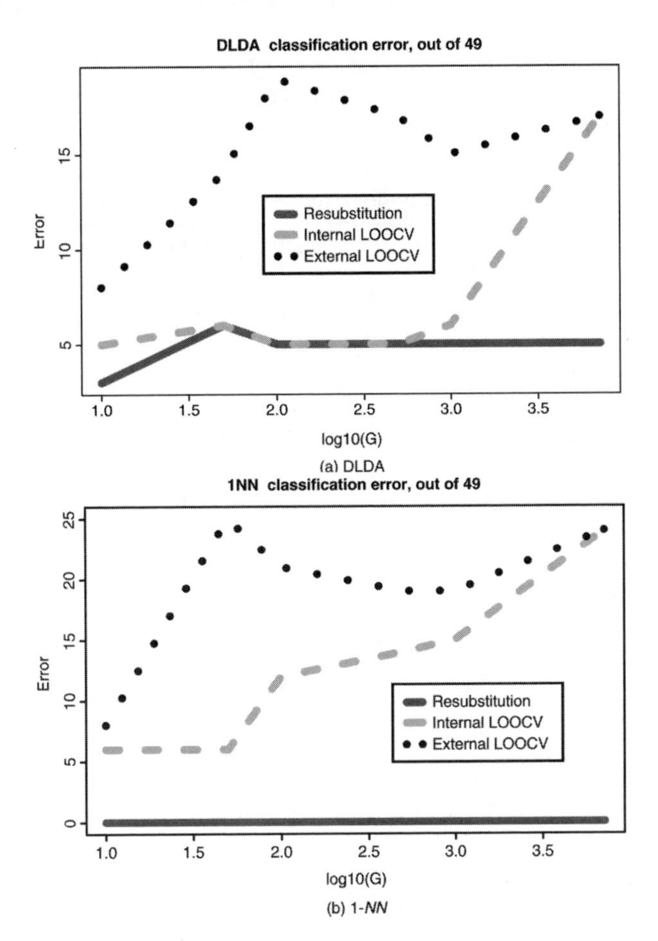

Figure 3.13 *(see color insert following page 114.) Breast tumor nodal dataset. Comparison of classification error estimation procedures: (solid lines) resubstitution estimation, where the entire learning set is used to perform feature selection, build the classifier, and estimate classification error; (dashed lines) internal LOOCV, where feature selection is done on the entire learning set and LOOCV is applied only to the classifier building process; (dotted lines) external LOOCV, where LOOCV is applied to the feature selection and the classifier building process. Error rates are plotted vs. the number of genes G (on log-scale) for (a) DLDA and (b) 1-NN.*

cross-validation study are reported for k-NN and naive Bayes classifiers in Table 3.12. These error rates are higher than those in Table 3.3 to Table 3.6. In addition, parameters estimated by LOOCV (inner LOOCV) tended to vary substantially, as observed with k for k-NN and the variance estimator for naive Bayes classified. LOOCV most often selected pooled variance estimators for naive Bayes classifiers and a small number of neighbors $k = 1$ or 3 for k-NN. Estimates of the number of genes were very variable for all but the nodal dataset, where G was always estimated as 10. For the two brain

Table 3.12 *Full cross-validation — estimates of classification error for k-NN and naive Bayes, using leave-one-out cross-validation to select classifier parameters and to estimate the overall error rate (i.e., double CV).*

	Breast tumor dataset		Brain tumor dataset	
	Estrogen	Nodal	MD classic vs. desmoplastic	MD survival
k-NN				
Euclidean distance				
Error rate	$6/49$	$9/49$	$5/34$	$25/60$
Number of neighbors, k	1 (60%)	1 (47%)	3 (70%)	3 (70%)
Number of genes, G	10 (71%)	10 (100%)	100 (56%)	500 (37%)
Naive Bayes				
Error rate	$6/49$	$9/49$	$7/34$	$24/60$
Variance estimator	pooled (65%)	pooled (91%)	pooled (80%)	pooled (56%)
Number of genes, G	50 (57%)	10 (100%)	50 (56%)	10 (55%)

Note: For k-NN, the number of genes $G = 10, 50, 100, 500, 1000$ and the number of neighbors $k = 1, 3, 5$ were selected by LOOCV. For naive Bayes, the number of genes $G = 10, 50, 100, 500, 1000$ and the type of variance estimator (pooled vs. unpooled, or DLDA vs. DQDA) were selected by LOOCV. The entries in parentheses indicate the percentage of the CV steps (out of n for LOOCV) for which the corresponding parameter values were selected. Variables were selected using a two-sample t-statistic.

tumor datasets, G was generally smaller for naive Bayes classifiers than for k-NN, while for the estrogen data G was generally smaller for k-NN.

3.8 Discussion

Classification is an important question in microarray experiments, for purposes of classifying biological samples and predicting clinical or other outcomes using gene expression data. Although classification is by no means a new subject in the statistical literature, the large and complex multivariate datasets generated by microarray experiments raise new methodological and computational challenges. These include building accurate classifiers in a "small n, large p" situation and obtaining honest estimates of classifier performance. A number of recent articles have addressed statistical issues related to classification in the context of microarray experiments: Ambroise and McLachlan (2002); Chow et al. (2001); Dettling and Bülmann (2002); Dudoit et al. (2002a); Golub et al. (1999); Moler et al. (2000); Pomeroy et al. (2002); Shieh et al. (2002); Tibshirani et al. (2002); West et al. (2001). These have mostly focused on existing methods or variants thereof, and, in general, comparison studies have been limited. So far, most articles on tumor classification have applied a single technique to a single gene expression dataset. Furthermore, comparison studies have not always been properly calibrated and, as reported by Ambroise and McLachlan (2002) and West et al. (2001), performance assessment procedures have often been inaccurate, due to the common practice of omitting gene screening from the validation process.

This chapter introduced the statistical foundations of classification and reviewed a number of commonly used classifiers. The important issues of feature selection and honest classifier performance assessment were discussed. A broad range of classifiers for predicting tumor class were evaluated using gene expression measures from studies of brain and breast cancer. The results confirmed earlier findings that simple methods, such as nearest neighbor and naive Bayes classification, are competitive with more complex approaches, such as aggregated classification trees or support vector machines (Dudoit et al., 2002a). These basic methods are intuitive and require little training; they are thus advisable and safer for inexperienced users. As learning sets increase in size, we may see, however, an improvement in the performance of more complex classifiers such as random forests. These methods may also be used to gain a better understanding of the predictive structure of the data and in particular gene interactions which may be related to class distinction. Other useful byproducts of random forests are vote margins and out-of-bag error rates. Although classifier performance did not always deteriorate with an increasing number of irrelevant features, some screening of the genes down to, for example, $G = 10 - 100$, appears to be advisable. Finally, we have demonstrated the importance of honest classifier performance assessment, which takes into account gene screening and other training decisions in error rate estimation procedures such as cross-validation.

Decisions that could affect classifier performance and were not examined here are the choices of class priors and loss function. In addition, we did not consider polychotomous vs. binary classification. These questions are most important in situations where the learning set comprises several unbalanced classes. The results of Dudoit et al. (2002a) for 3- and 8-class microarray datasets are consistent with those of the present chapter; however, a more complete study, considering class priors and loss function, is still needed. A number of other issues which were only briefly addressed here deserve further study. Similar to most articles published to date, we have focused on the prediction of polychotomous responses. However, in many important applications of microarray experiments, such as clinical trials, the outcome of interest may be quantitative (e.g., blood pressure, lipid level) or censored (e.g., censored survival time). In addition, more accurate predictions may be obtained by inclusion of other predictor variables such as age or sex. For the purpose of comparing classifiers, we have relied on naive one-gene-at-a-time screening procedures. Such univariate approaches have obvious limitations. The selected gene subsets may include a large number of genes with highly correlated expression measures, which provide redundant information on classification and may mask important genes with smaller main effects. In addition, such methods do not allow the detection of genes with weak main effects but strong interactions. Use of multivariate feature selection procedures may improve classifier performance and, more important, yield insight into the predictive structure of the data (i.e., identify individual genes and sets of interacting genes that are related to class distinction). Further investigation of the resulting genes may improve our understanding of the biological mechanisms underlying class distinction, and may eventually lead to the identification of marker genes to be used in a clinical setting for predicting outcomes such as survival and response to treatment. Although we have stressed the importance of accurate estimation of classification error and presented a number of different resampling procedures for estimating these error rates, the comparison study for the brain and breast tumor datasets relied primarily on leave-one-out cross-validation. A thorough study of the distributional properties of different error rate estimators is needed.

The clinical application of microarrays to cancer diagnosis and treatment requires the development of accurate classifiers, which provide confidence measures for individual predictions and are based on a subset of well-studied marker genes and clinical covariates. Much more complete studies of predictor performance and biological validation are needed before such an approach can be recommended. For microarray-based prediction methods to ever become useful and reliable tools in the clinic, it will be necessary to assemble comprehensive databases of tumor samples diagnosed by traditional means and with associated gene expression profiles and relevant clinical covariates. Designing clinical studies for this purpose will require the close collaboration of clinicians, biologists, and statisticians. The number of test samples needed to truly assess the clinical performance of any of the classifiers described in this chapter is much larger than what is currently available. Furthermore, these samples should come from diverse and well-defined sources (hospitals, labs), and should represent a variety of patients with different characteristics such as age, sex, race, type of tumor, etc.

The microarray classification studies performed to date have been based on small sample sizes and can be viewed only as preliminary and limited indications of the feasibility of microarray-based diagnosis and treatment. The following points can be made:

1. It is possible to use microarray expression measures to distinguish among different biological outcomes, such as tumor class. Due to limited sample sizes, however, it is not clear how accurately microarray expression measures can distinguish among different outcomes, especially those that reflect subtle biological differences.
2. None of the studies performed thus far have provided evidence in favor of a single best classifier.
3. With the currently available datasets, complex classifiers do not appear to improve upon simple ones.
4. The tumor classification error rates reported in the literature are generally biased downward, i.e., overestimate the accuracy with which biological and clinical outcomes can be predicted based on expression measures. Any realistic estimate of

generalization error (i.e., of the error rate for a new sample) should be based on large and representative test sets and should take into account all classifier training decisions (e.g., number of genes, classifier parameters such as the number of neighbors k in k-NN).

5. Tracking classifier parameter values in a full (double) cross-validation study can reveal important aspects of the data. Therefore, cross-validation should not only be viewed as an error rate estimation procedure, but also as an exploratory analysis tool.

In conclusion, before patients can benefit directly from the use of microarrays in cancer research, well-designed studies are needed to develop and refine accurate microarray-based predictors of clinical outcomes.

3.9 Software and datasets

Most classification procedures considered in this chapter are implemented in open source R packages (Ihaka and Gentleman, 1996), which may be downloaded from the Comprehensive R Archive Network (`cran.r-project.org`) or the Bioconductor Web site (`www.bioconductor.org`).

Linear and quadratic discriminant analysis. Linear and quadratic discriminant analyses are implemented in the `lda` and `qda` functions from the `MASS` package. DLDA and DQDA are implemented in the `stat.diag.da` function from the `sma` package. Mixture and flexible discriminant analysis procedures are implemented in the `mda` package.

Nearest neighbor classifiers. The `knn` function from the `class` package can be used for k-nearest neighbor classification.

Classification trees. CART of Breiman et al. (1984) are implemented in the commercial software package CART version 1.310 (California Statistical Software, Inc.) or the more recent CART 4.0 PRO (Salford Systems). They are also implemented in the R package `rpart`.

Bagging. The R package `ipred` implements direct and indirect bootstrap aggregation in classification and regression, and provides resampling-based estimators of prediction error.

Random forests. Fortran code and R interface for Random Forests version 1 may be downloaded from `oz.berkeley.edu/users/breiman`. Random forests are also implemented in the R package `randomForest`.

LogitBoost. The R package `LogitBoost` provides an implementation of the LogitBoost algorithm for tree stumps described in Dettling and Bülmann (2002).

Support vector machines. An implementation of SVMs is available in the `svm` function from the R package `e1071`.

The following Web sites provide access to a large number of microarray datasets: Stanford Genomic Resources (`genome-www.stanford.edu`) and Whitehead Institute, Center for Genome Research (`www-genome.wi.mit.edu/cancer`).

3.10 Acknowledgments

We are most grateful to Leo Breiman for many insightful conversations on classification. Marcel Dettling's help with the `LogitBoost` package and preprocessing of the breast tumor dataset was much appreciated. Finally, we acknowledge Robert Gentleman for valuable discussions on classification in microarray experiments while designing a short course on this topic.

CHAPTER 4

Clustering microarray data

Hugh Chipman, Trevor J. Hastie, and Robert Tibshirani

4.1 An example

We begin with an example that will be used throughout the chapter. The data come from Sørlie et al. (2001). The goal of that article was to "classify breast carcinomas based on variations in gene expression derived from complementary deoxyribonucleic acid (cDNA) microarrays and to correlate tumor characteristics to clinical outcome." The data consist of log fluorescence values for 456 cDNA clones measured on 85 tissue samples. Of the 85 samples, 4 are normal tissue samples, 78 are carcinomas, and 3 are fibroadenomas. Three of the four normal tissue samples were pooled normal breast samples from multiple individuals. Sørlie et al. (2001) selected the 456 genes from an initial set of 8102 genes so as to optimally identify the intrinsic characteristics of breast tumors. In Figure 4.1 and Figure 4.2, the data are plotted as heat maps.* This representation assigns a color for every matrix entry, with negative (underexpressed) values being green, and positive (overexpressed) values red. The data presented in this plot were preprocessed by Sørlie et al. (2001), adjusting rows and columns to have median zero. This preprocessing was applied before selection of the subset of 456 genes, so the column (i.e., sample) medians are not precisely zero.

Heat maps are used to look for similarities between genes and between samples. They are most effective if rows and columns are ordered so as to allow these patterns to be identified. Clustering is often used to give this ordering, by identifying groups of samples (genes) and then arranging the groups so that the closest groups are adjacent. This is illustrated in Figure 4.1, where rows and columns are arranged according to separate hierarchical clusterings. Sørlie et al. (2001) used a similar graphic to identify interesting groups of genes and tumor subtypes. In Figure 4.2, five interesting gene subgroups are given. These are similar to those identified by Sørlie et al. (2001). These gene groups were selected because of unusually high or low expression levels among some of the tumors (columns). The gene groups highlighted in Figure 4.2 are used to characterize the different tumor subtypes. The six tumor subtypes (indicated by color from left to right of the dendogram in Figure 4.2) are Basal-like (red),

* These plots were generated using Michael Eisen's `Cluster` and `Treeview` packages.

1-58488-327-8/03/$0.00+$1.50
© 2003 by CRC Press LLC

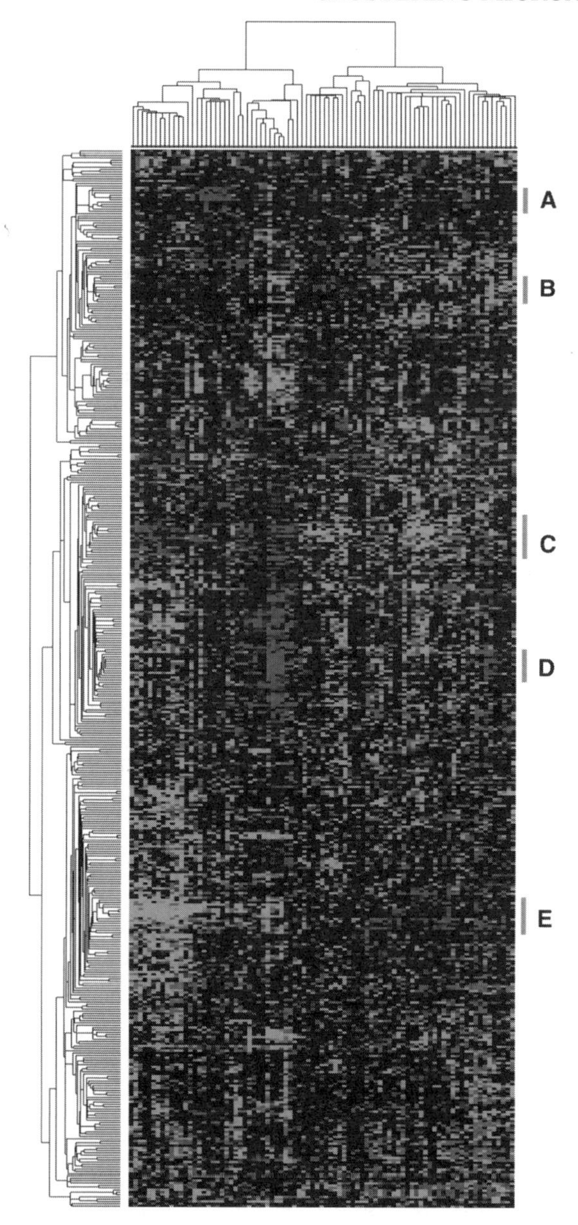

Figure 4.1 *(See color insert following page 114.) Heat map of the microarray data, with rows and columns arranged according to a hierarchical clustering method. Grey pixels represent missing data. Letters A–E and corresponding grey bars represent groups of genes displayed in greater detail in Figure 4.2.*

ERBB2+ (pink), Normal Breast-like (green), Luminal Subtypes C, B, and A (light blue, orange, dark blue). For example, the Basal-like subgroup (red) is characterized by high expression levels of genes in group C, and low levels in group E. The high

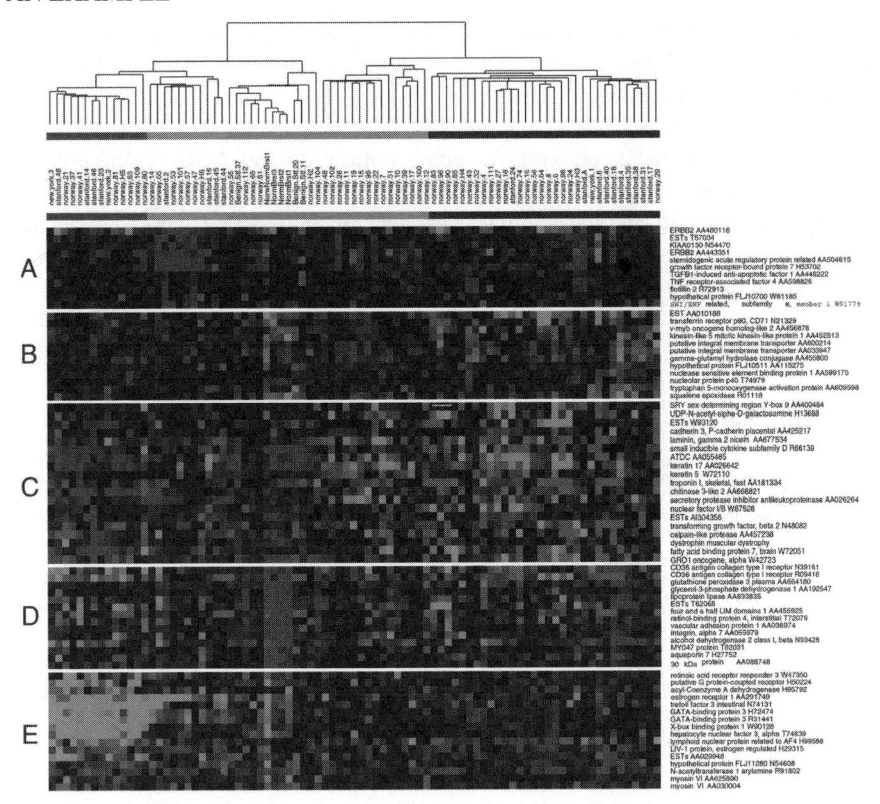

Figure 4.2 *(See color insert following page 114.) Enlarged segments of the heat map. Interesting groups of genes (rows) have been selected from Figure 4.1, as indicated by the letters A–E. Tumor subtypes corresponding to clusters of columns are indicated by color bars on the upper dendogram. See text for discussion.*

levels in group C include keratin 17, laminin, and fatty acid binding protein 7. Within the Normal Breast-like tumor group are all "Benign" and "Normal" samples. The four Normal samples are characterized by "the highest expression of many genes known to be expressed by adipose tissue and other nonepithelial genes" (group D).

Clustering methods are of interest beyond their applications to heat maps. Researchers wish to identify groups of samples that have similar expression level patterns, and genes that are similar across samples. Clustering may also be used for "supervised" learning tasks, looking at the association between clusters and some medical outcome. Although this is termed supervised learning, the outcome is not used in construction of the clusters.

In the remainder of this chapter, discussion of general methods is phrased in terms of clustering "objects." For microarray data, these objects could be either samples or genes. Most of our examples emphasize the clustering of samples.

The rest of the chapter is organized as follows: Section 4.2 discusses measures of dissimilarity and issues relating to missing data and imputation. An introduction to clustering methods is given in Section 4.3. Partitioning methods such as K-means, K-medoids, and self-organizing maps are discussed in Section 4.4. Section 4.5 covers a variety of hierarchical methods, including bottom-up and top-down algorithms. Two-way clustering methods are discussed in Section 4.6, and methods relating to principal components and the singular value decomposition are discussed in Section 4.7. Section 4.8 mentions other work in clustering microarray data, and software is discussed in Section 4.9.

4.2 Dissimilarity

Fundamental to clustering is a measure of similarity (or dissimilarity) of the objects being clustered. Clustering seeks to group together those objects which are most similar (or least dissimilar). The within-cluster dissimilarity can form the basis for a loss function that clustering seeks to minimize, in much the same way that linear regression seeks to minimize the sum of squared distances between observed and fitted values.

Common measures of dissimilarity for continuous data include Euclidean distance,

$$||x - y|| = \sqrt{\sum_{i=1}^{p}(x_i - y_i)^2}, \tag{4.1}$$

or its square,

$$||x - y||^2 = \sum_{i=1}^{p}(x_i - y_i)^2. \tag{4.2}$$

Here, x and y are p-vectors of measurements on the objects to be clustered. Also used are L_1 (Manhattan) distance,

$$d_{xy} = \sum_{i=1}^{p}|x_i - y_i|, \tag{4.3}$$

and the "$1 - correlation$" distance,

$$d_{xy} = 1 - \rho_{xy} = 1 - \frac{\sum_{i=1}^{p}(x_i - \overline{x})(y_i - \overline{y})}{\left[\sum_{i=1}^{p}(x_i - \overline{x})^2\right]^{1/2}\left[\sum_{i=1}^{p}(y_i - \overline{y})^2\right]^{1/2}}. \tag{4.4}$$

Note that the sums are over the p variables in x and y. The $1 - correlation$ distance is bounded in $[0, 2]$ (objects with correlations 1.0 and -1.0, respectively). This dissimilarity is invariant to changes in location or scale of either x or y. Variations on this distance include an uncentered version (replacing $\overline{x}$ and $\overline{y}$ with 0) and a version that uses the absolute value of correlation, giving $1 - |\rho_{xy}|$.

The $1 - correlation$ dissimilarity can be related to the more familiar Euclidean distance. If

$$\widetilde{x} = \frac{x - \overline{x}}{\sqrt{\sum_i (x_i - \overline{x})^2 / p}} \quad \text{and} \quad \widetilde{y} = \frac{y - \overline{y}}{\sqrt{\sum_i (y_i - \overline{y})^2 / p}},$$

then

$$||\widetilde{x} - \widetilde{y}||^2 = 2p(1 - \rho_{xy}).$$

That is, squared Euclidean distance for standardized objects is proportional to the correlation of the original objects.

The choice of a dissimilarity measure is best motivated by subject matter considerations, and microarray data are no exception. The location/scale invariance of the $1 - correlation$ dissimilarity makes it a popular choice for microarray data. Changes in the average measurement level or range of measurement from one sample to the next are effectively removed by this dissimilarity.

For the breast carcinoma data, we plot in Figure 4.3 the 456 expression values for the most and least similar pairs of samples. The least similar (right) have a correlation of $-.396$, while the most similar have a correlation of .850. The two most similar samples (NormBrst1, NormBrst2) are two of the three composites of normal breast tissues included in the study. Because expressed genes were selected to emphasize carcinoma, it appears plausible that regular tissue samples would appear more similar to each other.

To aid comparison of different clustering algorithms in the breast carcinoma example, we elect to use squared Euclidean distance applied to scaled variables (i.e., $||\widetilde{x} - \widetilde{y}||^2$).

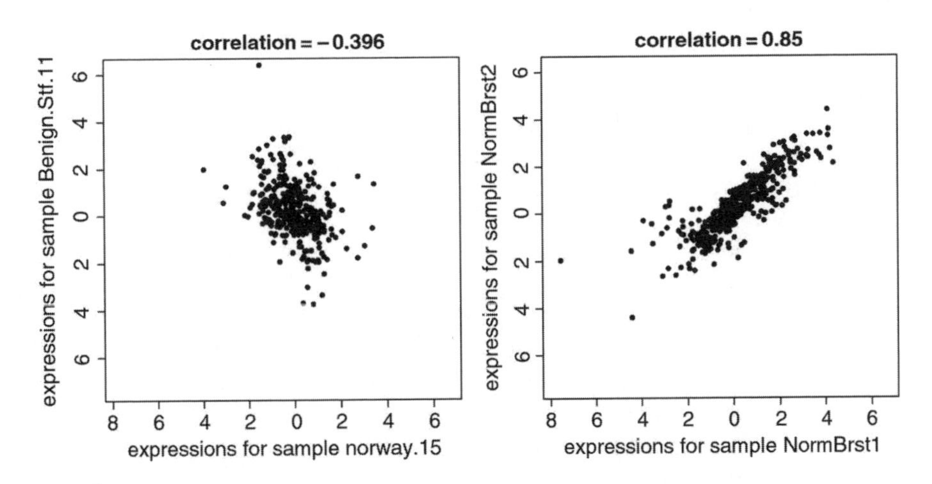

Figure 4.3 *Expression levels for the least (left) and most (right) similar pairs of tissue samples, according to the 1-correlation dissimilarity. Correlation values are given above each plot.*

4.2.1 Missing data and imputation

Missing data are quite common in microarray data. In the breast carcinoma example, almost 6% of values are missing. This missingness is not uniform across samples: some samples have no missing values, while others have up to one-third missing.

Two strategies for dealing with missing values are either to modify clustering methods so that they can deal with missing values, or impute a "complete" dataset before clustering. Each of these strategies are described below. The examples in this chapter will use imputed data. This decision was made because of the comparative aspect of this chapter. In practice, if methods that were robust to missing values were to be used, imputation would be unnecessary.

We first describe modification of methods to deal with incomplete data. The easiest modifications are for clustering methods that use only a dissimilarity matrix. These include K-medoids, bottom-up clustering and some forms of top-down clustering, described in Section 4.5.1 and Section 4.5.2. To calculate a dissimilarity between two objects with some missing values, a pairwise deletion strategy is often used. Such pairwise deletion is illustrated for a subset of the data, given in Table 4.1 (10 genes, 4 samples). To calculate the dissimilarity between samples BC48.0 and BC119A.BE, genes 1, 3, 8, and 9 would be excluded because they have one or more missing values in these samples. For comparing samples BC48.0 and BC205B.AF, genes 1, 8, and 9 would be excluded. Dissimilarity measures such as Euclidean (Equation 4.1) or Manhattan (Equation 4.3) that are proportional to the number of terms in the summation, require adjustment for the varying number of complete terms used. If c of the p

Table 4.1 *A subsection of the expression data (used in the text to illustrate treatment of missing values).*

	Sample			
Gene	BC790	BC48.0	BC119A.BE	BC205B.AF
1	−0.34	—	0.01	0.79
2	0.23	−0.88	−1.49	−0.59
3	−0.49	−0.32	—	−0.03
4	0.05	3.05	−1.01	−1.17
5	0.38	3.20	−2.74	−2.65
6	0.70	2.91	−4.67	−4.55
7	0.56	3.02	−1.55	−2.08
8	0.75	—	−1.65	—
9	−0.56	—	−0.40	−0.75
10	−0.69	1.33	−2.21	0.44

measurements are complete for both x and y, then a multiplier p/c should be included inside the summation. Such an adjustment is not required for the $1 - correlation$ dissimilarity because it would cancel out of the numerator and denominator.

Missing values may also affect the equivalence of $1 - correlation$ and squared Euclidean distance, depending on how these quantities are calculated. Pairwise deletion appears to be the most natural way to calculate a correlation. For a squared Euclidean distance, the data would typically be scaled once before calculating Euclidean distance; pairwise deletion would be used only when calculating distance between the scaled data points. The scaling implicit in correlation uses different subsets, while the scaling for Euclidean distance always uses the same subsets. For moderate amounts of missing data, this difference can be minor, as indicated by a plot of squared scaled Euclidean distance against $1 - correlation$ distance for the breast carcinoma data in Figure 4.4.

Clustering algorithms that do not rely only on distances need further modification for missing values. Some methods, such as the bottom-up clustering of Eisen et al. (1998), have particular ways of dealing with missing values. These are discussed in Section 4.5.1.

Imputation is a commonly used method for dealing with incomplete data, especially in conjunction with some methods (e.g., K-means) that are not as easily adapted to missing values. The simplest method of imputation replaces missing values with the mean level for that gene. Better results can be obtained by using dependencies between different genes. For example, in Table 4.1, to replace the missing values for genes 1, 8, and 9 of BC48.0, a predictive model with BC48.0 as the response and the other three variables as predictors could be trained using observations 2–7, and 10. This model is then used to predict the missing values in BC48.0. Depending on the type of predictive model, complete data may also be required on the predictor

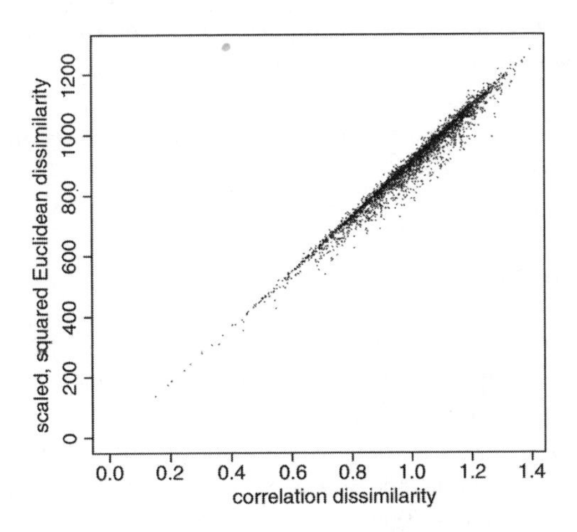

Figure 4.4 *Comparison of pairwise dissimilarities, using correlation distance and squared Euclidean distance applied to scaled samples.*

variables (here, BC790, BC119A.BE, and BC205B.AF), which would mean that only rows 2, 4–7, and 10 could be used to train the model. The large size of the data matrix in microarray applications means that there may be very few complete rows of data. In the breast carcinoma data, 84% of rows (genes) are incomplete.

Three training methods that are able to deal with incomplete data in the predictor variables are described in Troyanskaya et al. (2001) and Hastie et al. (1999). In the first, which uses a singular value decomposition (SVD), initial guesses are made for missing values, and then a reduced-rank SVD of the data is calculated. This SVD is used to generate new guesses of the missing values, and another SVD is calculated. This continues until a steady-state is reached. Although convergence is not guaranteed, this algorithm does have as a fixed point a solution to a least squares prediction problem for the missing data. The second approach uses K-nearest neighbors to impute missing data. Missing values among the predictors are not used in determining the nearest neighbors. The third strategy is to iteratively predict missing values for each variable using a regression model in which all other variables are predictors. In principle, any regression model (linear regression, K-nearest neighbors, trees, etc.) might be used. The iterative nature of this procedure arises because the most recently imputed values of predictors are used to fill in the missing values of the response.

Typically, a tuning parameter is used in all the imputation strategies mentioned previously (the rank of SVD used, the number of neighbors, etc.). Troyanskaya et al. (2001) and Hastie et al. (1999) examine how performance varies with these parameters, and suggest using about ten nearest neighbors, and rank 7 or less for the SVD. These recommendations are based on experiments where additional values were deleted so that predictions could be compared to actual values.

An additional technique for determining accuracy of different methods is to compute distances for the imputed data and compare these to robust distances calculated with the unmodified data matrix. This comparison could be made either among genes or among samples. In Figure 4.5, we plot imputed versus robust distances between genes (top row) and samples (bottom row). These plots are made for both the SVD method (with rank 7) and a knn-based regression strategy (with $K = 10$). We see that the knn strategy comes much closer to the robust distances, and consequently will use knn-imputed data in the remainder of the chapter.

4.3 Clustering methods

Clustering algorithms seek to assign N observations in p-space labeled $x_1, x_2, \ldots, x_N$ to one of K groups. This assignment operation can be characterized as a many to one mapping from observations to clusters, with an encoder function C. That is, observation i is mapped to cluster $C(i)$. Clustering algorithms seek to identify an optimal encoder C^*. For example, an encoder that minimizes total with-cluster dissimilarity might be desired. Most methods assume that each point belongs to one and only one cluster. Underlying probability models for the data are often ignored, although we will note connections to probability models in subsequent sections.

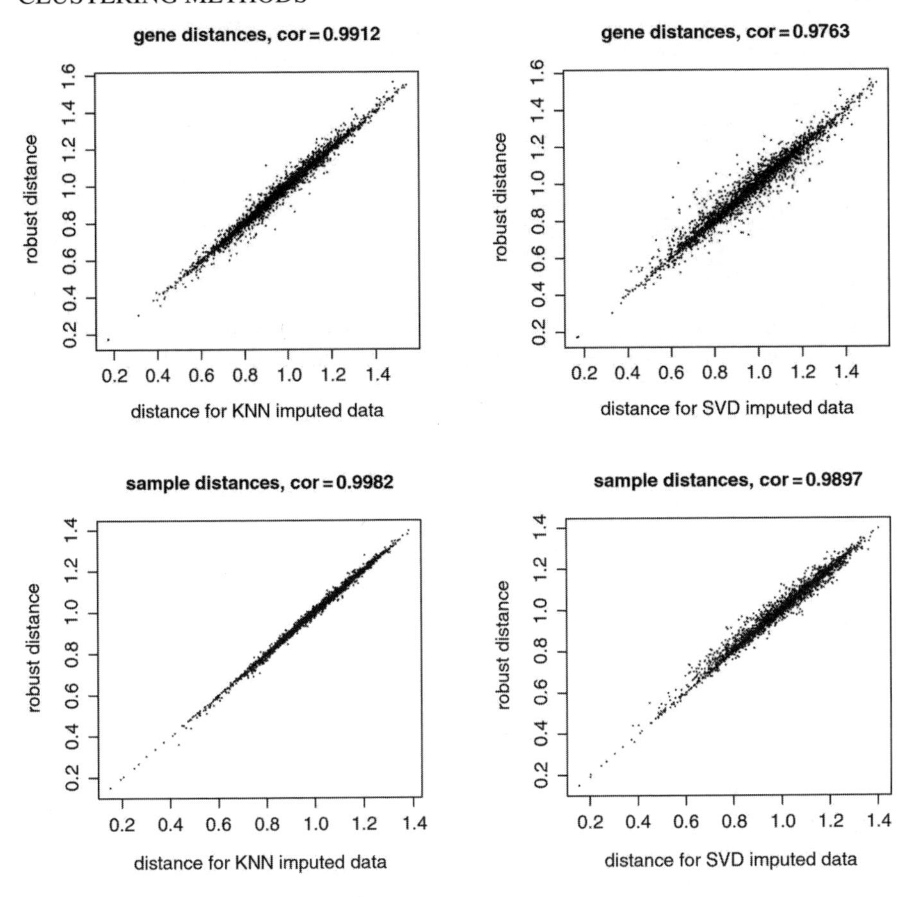

Figure 4.5 *Each scatterplot is of pairwise distances based on imputed data against corresponding robust distances based on original data. The top row corresponds to distances between genes, the bottom between samples. Left and right plots correspond to two different imputation schemes.*

Many clustering methods can be broadly divided into two classes: *partitioning methods*, which seek to optimally divide objects into a fixed number of clusters, and *hierarchical methods*, which produce a nested sequence of clusters.

Before pursuing details, we give a simple example to illustrate concepts and contrast hierarchical and partitioning methods. On the left of Figure 4.6, six points are plotted in two dimensions. The polygons surrounding them indicate a hierarchical clustering, which is also illustrated with a *dendogram* on the right side of the figure. In this case, a bottom-up algorithm was used to merge successive observations, starting with the two closest points 1 and 6. Nearest clusters are merged until all points are in one cluster. Once points are joined, they are never subdivided, giving a nested hierarchical clustering.

Figure 4.7 plots different partitions of the same data into 2, 3, 4, and 5 clusters by K-means, a partitioning algorithm. For 4 or 5 clusters, the points are grouped in the

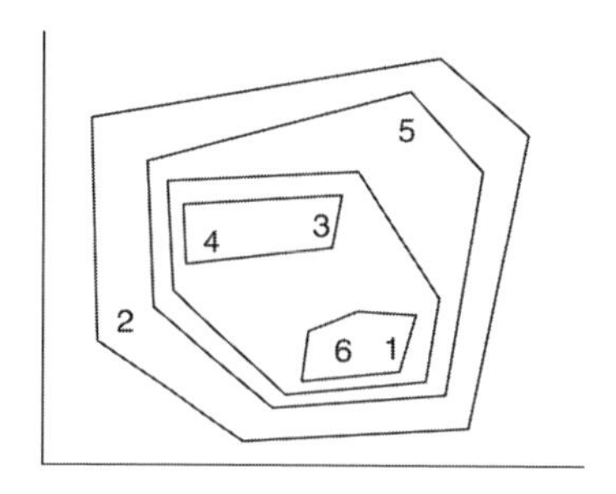

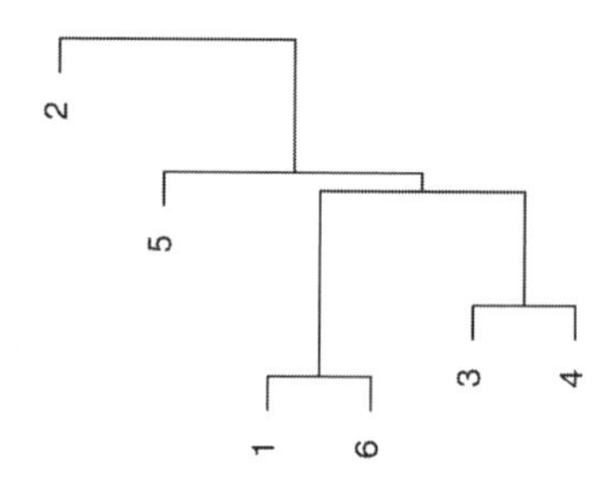

Figure 4.6 *Hierarchical clustering example. Points in two-dimensional space are illustrated on the left, and the corresponding hierarchical clustering is depicted on the right.*

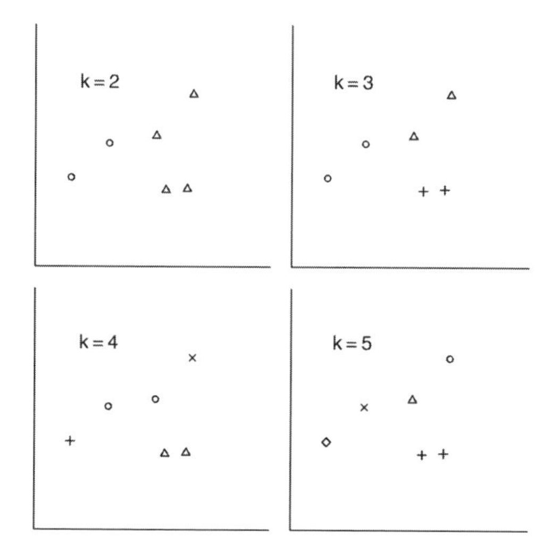

Figure 4.7 *K-means clustering example, with $k = 2-5$ clusters. Cluster membership is indicated by plotting symbol.*

same way as with hierarchical clustering. For 3 clusters, however, the two points in the center of the plot that were joined in the 4-cluster case are now separated again: successive clusterings are not nested.

Details of the preceding clustering methods are given in the next two sections.

4.4 Partitioning methods

Partitioning methods seek to minimize some measure of within-group dissimilarity for a fixed number (K) of groups. This is a combinatorial optimization problem, meaning that in most problems the global optimum will not be found, and one of possibly many local optima will be instead identified.

4.4.1 K-means

The most popular partitioning method, K-means, begins with K initial centers $\hat{\mu}_1$, $\hat{\mu}_2, \ldots, \hat{\mu}_K$, and alternates between mapping observations to the nearest center, and then averaging points within clusters to update centers. Algorithm 1 gives details. This method seeks to minimize the sum of squared distances from each observation to its cluster center $\hat{\mu}_k$,

$$WSS = \sum_{k=1}^{K} \sum_{C(i)=k} ||x_i - \hat{\mu}_k||^2. \tag{4.5}$$

Algorithm 1 *K-means clustering.*

1. For a given cluster assignment C, the total cluster variance (Equation 4.5) is minimized with respect to $\{\hat{\mu}_1, \ldots, \hat{\mu}_K\}$ yielding the means of the currently assigned clusters.
2. Given a current set of means $\{\hat{\mu}_1, \ldots, \hat{\mu}_K\}$, (Equation 4.5) is minimized by assigning each observation to the closest (current) cluster mean. That is,
$$C(i) = \text{argmin}_{1 \leq k \leq K} ||x_i - \hat{\mu}_k||^2.$$
3. Steps 1 and 2 are iterated until the assignments do not change.

For complete data, a modified form of Equation (4.5) is related to the sum of squared distances between points in each cluster,

$$\sum_{k=1}^{K} n_k \sum_{C(i)=k} ||x_i - \hat{\mu}_k||^2 = \frac{1}{2} \sum_{k=1}^{K} \sum_{C(i)=k} \sum_{C'(i)=k} ||x_i - x_{i'}||^2. \tag{4.6}$$

Here, cluster k has n_k points.

Different solutions will be achieved for different starting values; as a consequence, it is good practice to use multiple runs of the algorithm. In the breast carcinoma examples, ten restarts were used for each value of K in K-means and tree structured vector quantization (TSVQ) (TSVQ is an algorithm built upon K-means; see Section 4.5.2). Also, Algorithm 1 does not guarantee a optimal local minimum, as it may be possible to reassign points to different clusters and further reduce sums of squares. Hartigan and Wong (1979) give a more complicated algorithm which ensures that there is no single switch of an observation from one group to another group that decreases (Equation 4.5).

The K-means algorithm is fast, as it never evaluates all $N(N-1)/2$ pairwise dissimilarities. At each iteration, KN dissimilarities are evaluated, and K centroids updated. This speed makes K-means a popular algorithm, allowing it to cluster thousands of objects.

In principle, the K-means algorithm should have little difficulty with missing data because mean updates and distance calculations (steps 1 and 2 in Algorithm 1) can be performed with some missing values. In practice, fast update formulas in K-means and in related algorithms such as that of Hartigan and Wong are more difficult to modify for missing values. For this algorithm, it will be easier to use imputed data.

In addition to being robust to missing values, clustering algorithms may be affected by outliers. Because K-means minimizes a sum of squared Euclidean distances, it may be more sensitive to large distances than methods that minimize Euclidean distance. Such methods are described in the next section.

A connection exists between the K-means algorithm and mixture models for multivariate normal distributions. Minimizing within-cluster squared distance to a centroid is equivalent to maximizing the likelihood for a mixture of normals with spherical covariance structure. The K-means algorithm is also a close analogue of the expectation maximization (EM) algorithm used to estimate mixture models. Thus, mixtures of normals can be thought of as a "soft" version of K-means, with membership probabilities replacing the "hard" encoder function C. With identity covariance, mixtures of normals and K-means agree closely. One advantage of mixture models is that the probability models can be used in criteria to choose an appropriate number of clusters. Results become substantially different (and mixture modeling much more interesting) when the covariance structure of the multivariate normal distributions composing the mixture is relaxed. Estimation of such covariance structure is only possible for large numbers of points in small to moderate-dimensional spaces. Consequently, mixture models might be more appealing for clustering genes than samples. Indeed, Yeung et al. (2001) use mixture models to cluster genes. An important issue is appropriate data transformations to achieve normality of the data. The article also reviews theory and methods in mixture model clustering methods.

An important practical issue for partitioning methods is how to choose an appropriate number of clusters. Typically, a partitioning method is run repeatedly for different K, and a loss measure plotted against the number of clusters. These loss measures are not usually motivated by inferential considerations, due to the lack of explicit probability models for the data.

Figure 4.8 illustrates one loss measure—the within-cluster sum of squared Euclidean distance

$$\sum_{k=1}^{K} \sum_{C(i)=k} \sum_{C(j)=k} d_{ij}/2n_k - \sum_{k=1}^{K} \sum_{C(i)=k} ||x_i - \overline{x}||^2.$$

This measure is plotted against cluster size for K-means and a variety of methods discussed later. To accentuate differences, the right plot gives the value above the minimum for each cluster size. K-means has the smallest within-group distance, for $K \leq 50$, so it is, in some sense, finding the "best" clusters. As the number of clusters (K) increases, the algorithm may get trapped in more local optima. All the other methods are more constrained than K-means in one way or another, such as producing nested sequences of clusters or requiring that each cluster center be an observed data point.

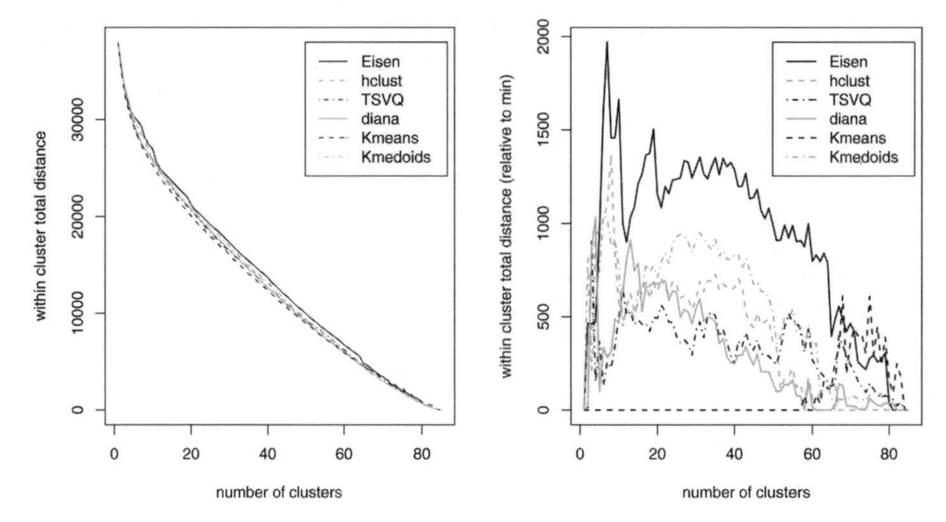

Figure 4.8 *Comparison of within-group dissimilarities (left) for various clustering methods, as functions of cluster size. The right plot gives difference from the minimum.*

A plot of within-cluster sum of distances, such as Figure 4.8, can be used to select a number of clusters, by looking for an "elbow." Tibshirani et al. (2001) refine this idea with the gap test, making a comparison between the change in within-cluster dispersion and an expected change under an appropriate reference null distribution. See Section 4.7.3, for details.

For illustrative purposes here, we select $K = 5$ clusters, the same number used in Sørlie et al. (2001). Various summaries of each cluster are possible, such as a heatmap with cluster centroids replacing samples, or numeric and graphical summaries of the clusters. In the breast carcinoma data, each sample has additional information, such as the age of the patient, survival time in months, survival status, etc. In Figure 4.9, a scatterplot of age against survival time is overlaid with the median age and survival time within each of the five clusters, for K-means and K-medoids (discussed in Section 4.4.2). The within-cluster medians for K-means are represented by diamonds. Three distinct groups are evident: one group of two clusters with a short survival time ($<$20 months), and two groups with longer survival times ($\approx$40 months). The longer survival time groups have median ages of around 55 (two clusters) and 65 (one cluster). Of course, the variables plotted were not used in the clustering. Instead of explaining why the clusters are separate, such a plot provides additional information about the identified clusters. The K-medoids groups in this plot are discussed in Section 4.4.2.

Clustering or classification?

As described previously, the approach of looking at auxiliary measures within each cluster can be extended to accomplish classification and regression. If a clinical

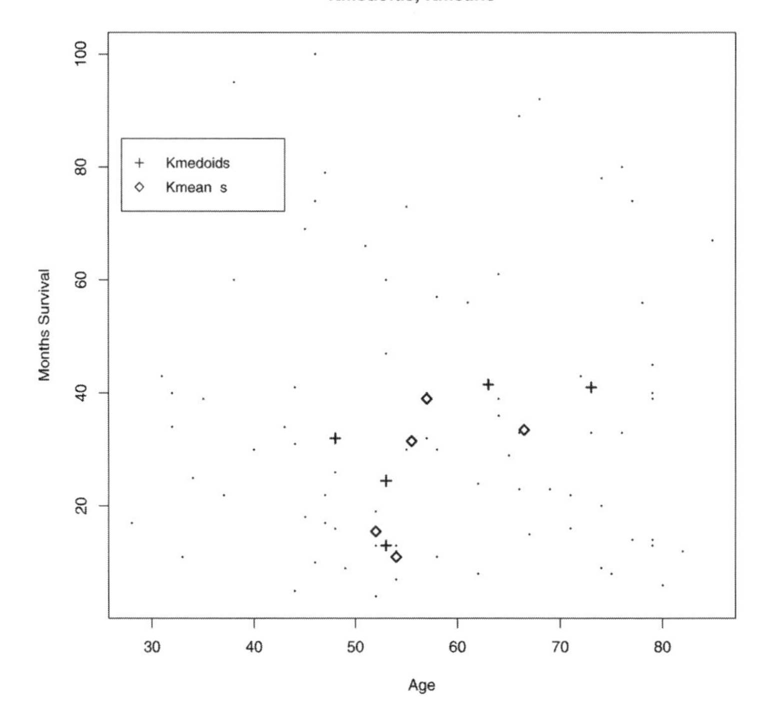

Figure 4.9 *Within-cluster medians of age and months of survival for six clustering methods explored in the chapter, and also for the clustering identified in Sørlie et al.*

outcome is to be predicted, sometimes clusters identified without knowledge of the outcome are used to predict the outcome. Whether or not this produces good predictions will depend on the extent to which the variables used in the clustering correspond to different outcome values. One possible advantage of such an approach is that it should be more resistant to overfitting because the response variable is not used in training (i.e., in constructing clusters). It does appear unlikely, however, to outperform an intelligently used classification or regression model.

4.4.2 K-medoids

The K-means algorithm takes averages of points assigned to each cluster to define cluster centroids. Such a centroid may have little interpretative value in some problems, such as when some variables are categorical or discrete. In these situations, it may be more meaningful for each cluster centroid to be a representative object (i.e., one of the observed data points). The *medoid* of a cluster of points is the point with smallest average dissimilarity to all other points. Thus, the K-medoid algorithm is obtained by modifying step 1 of Algorithm 1 so that the medoid of each cluster is calculated instead of the mean, and step 2 is modified so that distance instead of squared Euclidean distance is used. This calculation is more expensive, requiring all pairwise dissimilarities within each cluster. With this modification, the K-medoids algorithm uses only the dissimilarity matrix and not the original data.

K-medoids can be robust to missing values and outliers. Use of the dissimilarity matrix instead of data matrix means only a dissimilarity measure that deals with missing values is required. Use of a medoid instead of a mean is more robust to outliers in a similar way that a univariate median is more robust than a mean.

A variety of other combinatorial algorithms have been proposed to solve the same problem as K-medoids. Kaufman and Rousseeuw (1990) propose a partitioning around medoids (PAM) algorithm with a similar flavor to that of Hartigan and Wong (1979). The algorithm first finds an initial set of medoids and then swaps points so that no single switch of an observation with a medoid will decrease the objective.

When applied to the breast carcinoma data (using squared Euclidean distance), K-medoids finds somewhat similar solutions to K-means. Figure 4.8 indicates that the within-cluster dissimilarities are slightly larger than K-means for most values of K. This may be due to the more constrained nature of the centroids.

For a choice five clusters, as in the K-means example earlier, the within-cluster medians of age and survival time are plotted in Figure 4.9. Three of the five clusters have age/survival medians moderately close to K-means clusters, with the remaining two K-medoids clusters differing most in terms of age.

4.4.3 Self-organizing maps (SOMs)

SOMs (Kohonen, 1990) are partitioning algorithms that are constrained so that clusters may be represented in a regular, low-dimensional structure, such as a grid. This facilitates graphical display: clusters that are close to one another appear in adjacent cells of the grid. Each of K clusters is represented by a prototype object $M_i, i = 1, \ldots, K$. The prototypes are points in the same space as the data, but the estimation algorithm constrains the prototypes to a low-dimensional, grid-like structure.

An example is given in Figure 4.10, with five clusters in two dimensions and corresponding prototypes arranged in a one-dimensional grid. The data are plotted on the left along with the prototypes; the grid layout of prototypes is on the right. The prototype coordinates in data space are like a distortion of the grid space so that the prototypes lie close to the clusters, while retaining a grid structure. Typically, SOMs would be used to reduce high dimensional data to a one or two-dimensional grid of clusters.

The SOM clustering algorithm is quite similar to K-means, but with a constraint reflecting prototype configuration. Algorithm 2 gives an online version of this clustering algorithm for the case of a two-dimensional grid. For two dimensions, a double

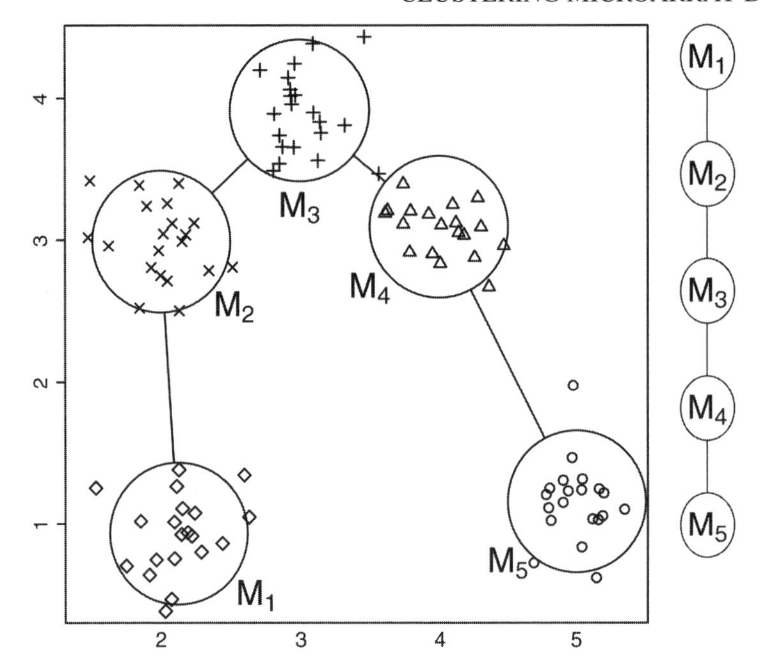

Figure 4.10 *Self-organizing map example. Points in two-dimensional space are clustered with a SOM configured to have five clusters on a one-dimensional grid. The ordering of clusters along the grid is indicated by the connected circles. The grid structure is indicated on the right.*

indexing scheme of prototypes in the grid space (by row and column) is convenient. Each step of the algorithm adjusts prototype coordinates using only one of the data points. The grid constraint is enforced by updates that move not just one prototype toward a data point, but also neighbors (in the grid space) of the nearest prototype. Such an algorithm would typically be run for several thousand iterations. Initial values of the grid radius r would depend on the number of clusters, but might be chosen so that about a third of all prototypes belong to the same neighborhood.

This algorithm is quite similar to K-means. In fact, if the grid radius r is small enough that only one prototype is updated in step (4c), then the SOM algorithm is equivalent to an online version of K-means. Use of such a small value of r means that all spatial constraints on the prototypes are lost.

The appeal of this algorithm is that the arrangements of prototypes on a grid allow neighboring clusters to be interpreted as being more similar. Tamayo et al. (1999) use this idea in the context of microarray data, with a two-dimensional grid. They present a grid-structured summary of the cluster represented by each prototype. Each summary is typically a plot of expression levels of a prototype gene across the different (time-ordered) experiments.

Algorithm 2 *SOM clustering (two-dimensional rectangular grid).*

1. Select the number of rows (q_1) and columns (q_2) in the grid. Then there will be $K = q_1 q_2$ clusters.
2. Initialize step size $\alpha = 1$ and grid radius $r = 2$ (for example).
3. Initialize prototype vectors $M_{\mathbf{j}}, \mathbf{j} \in (1, \ldots, q_1) \times (1, \ldots, q_2)$ by assigning points to prototypes. This assignment might be based on a partitioning of the data after projection onto principal components.
4. Loop over the entire dataset
   - Loop over each data point x_i
     (a) Identify the index vector $\mathbf{j}^*$ of the prototype $M_{\mathbf{j}}$ nearest to x_i.
     (b) Identify a set S of neighboring prototypes of $M_{\mathbf{j}^*}$, i.e.,
     $$S = \{\mathbf{j} : \text{distance}(\mathbf{j}, \mathbf{j}^*) < r\}.$$
     The distance might be Euclidean or some other metric.
     (c) Update each element of S by moving the corresponding prototype toward x_i:
     $$M_{\mathbf{j}} \leftarrow M_{\mathbf{j}} + \alpha(x_i - M_{\mathbf{j}}) \qquad \forall \mathbf{j} \in S$$
   - Decrease the values of α and r by some predetermined amount. Typically $\alpha = 0, r = 1$ upon completion of the outer loop.

One-dimensional SOMs also have interesting uses for microarray data. For example, Eisen's (1998) `Cluster` package can use a SOM with a one-dimensional grid of prototypes to aid the ordering of a hierarchical clustering dendogram. An algorithm is used to make the ordering of nodes of the dendogram resemble as closely as possible the one-dimensional SOM. The issue of reordering nodes of a dendogram is further discussed in Section 4.5.4.

4.5 Hierarchical methods

The nested sequence of clusters produced by hierarchical methods makes them appealing when different levels of detail are of interest because small clusters are nested inside larger ones. In microarray applications, interest may focus on both small groups of similar observations and a few large clusters. The former might occur when individuals provide multiple samples or a few samples have special meaning, such as the four samples in the breast carcinoma example that are normal tissue. The latter would occur when larger groups exist, such as samples from two different sources, or different stages of carcinoma, or from different experiments (the yeast data in Section 4.6.3 has this property).

Many hierarchical clustering methods have an appealing property that the nested sequence of clusters can be graphically represented with a tree, called a *dendogram* (e.g., Figure 4.6). Usually, each join in a dendogram is plotted at a height equal to the dissimilarity between the two clusters which are joined. Selection of K clusters from a hierarchical clustering corresponds to cutting the dendogram with a horizontal line at an appropriate height. Each branch cut by the horizontal line corresponds to a cluster. In some applications, clusters are identified that correspond to no horizontal cut. For example, in Sørlie et al. (2001) and Figure 4.2, the six tumor subgroups (clusters) indicated by color across the top of the plot do not correspond to a precise horizontal cut.

Although highly interpretable, dendograms can mislead. Figure 4.6 suggests that the (reasonably close) observations 2 and 4 are quite distant because they are not joined until the top of the dendogram. The hierarchical structure represented in a dendogram is imposed on the data by the clustering algorithm and will reflect actual patterns in the data only to the extent that these patterns are of a hierarchical nature.

To give a sense of the variety of dendograms produced by different clustering algorithms, small, unlabeled dendograms are plotted in Figure 4.11. Their appearances are affected by differing measures of within- and between-cluster dissimilarity, as seen in the vertical axes. More detailed descriptions of the trees are given in subsequent sections.

4.5.1 Bottom-up methods

The most common hierarchical methods are bottom-up, starting with each object forming a cluster of size 1. At each step, the closest two clusters are joined until all objects are in a single cluster, as in Figure 4.6. The measure of "closeness" has many possible definitions when clusters are not singleton points. These include:

Single linkage, which uses the minimum distance between points in different clusters

Complete linkage, which uses the maximum distance

Mean linkage, which uses the average of all distances between points in the two clusters

Distance between centroids, which represents each cluster by a centroid and measures inter-cluster distances using centroids

These are illustrated in Figure 4.12. Single linkage tends to produce long chains of clusters. It also is a solution to the minimum spanning tree problem. Complete linkage tends to produce compact, spherical clusters, and mean linkage is a compromise between the two. Other variations also exist, such as trying to minimize within-cluster variance, in a manner similar to K-means. This latter approach minimizes a total instead of a mean, so clusters with fewer observations are likely to be joined first. Mean linkage does not depend on the number of observations in each cluster. Dendograms labeled "Eisen" and "hclust" in Figure 4.11 use distance between centroids and mean linkage, respectively.

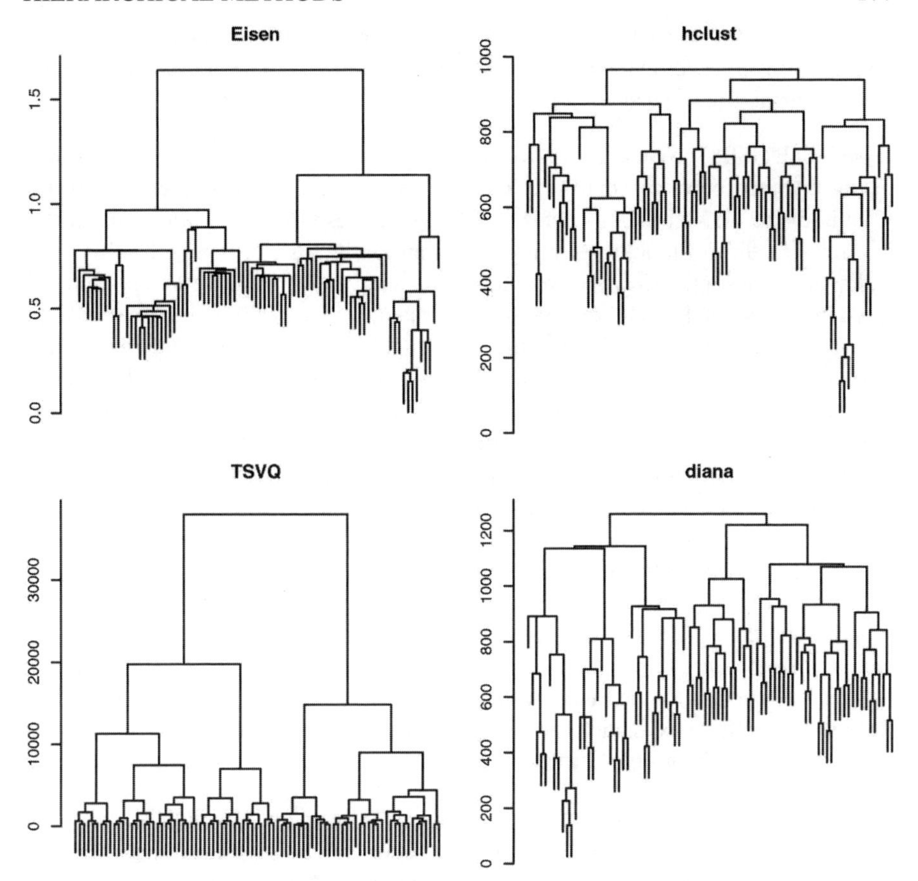

Figure 4.11 *Trees representing four different clustering algorithms. Differences in vertical scale and appearance are due to the use of different criteria. Eisen and hclust algorithms are bottom-up. TSVQ and diana are top-downs.*

Dendograms for the bottom-up methods using single, complete, and mean linkage possess a monotonicity property: the dissimilarity between clusters is monotone increasing with successive joins.

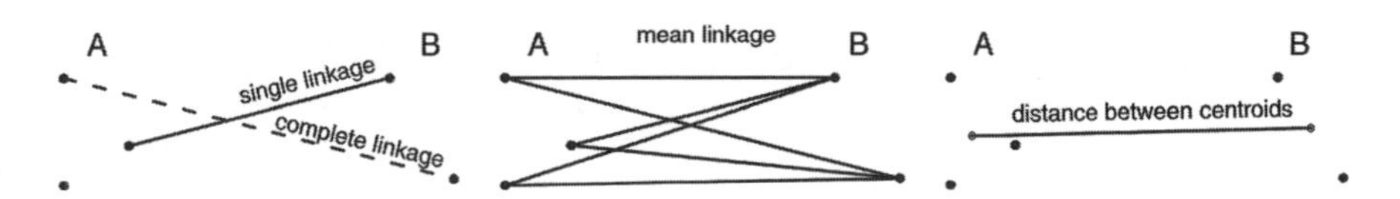

Figure 4.12 *Single linkage, complete linkage (left plot), mean linkage (middle plot), and distance between centroids (right plot) for two clusters A and B.*

Hierarchical clustering operates on the dissimilarity matrix instead of directly on observations. It is computationally expensive for data with many observations (large N), requiring $O(N^2)$ calculations.

Eisen et al. (1998) propose a variation on bottom-up, mean linkage clustering, instead using distance between centroids. They define centroids as within-cluster means. Centroids may have missing values for some variables if all observations in a cluster are missing those variables. Missing values are omitted in a pairwise fashion when calculating dissimilarities, and when updating a mean, all complete observations are used in each variable. In the current implementation (Eisen, 1998), the updating algorithm for centroids treats missing values in an unexpected manner. When two clusters are merged, the new centroid is calculated as a weighted average of the two cluster centers. If different variables have differing numbers of incomplete observations, this will not give the same center as recalculating the mean using all original observations. Although this algorithm usually gives similar results to average-linkage clustering, results can differ. Also, the resultant dendogram may be non-monotone, due to changing cluster centroids. In the breast carcinoma example, this problem occurs at 13 of the 84 distinct joins of the tree. The `Cluster` software implementation (Eisen, 1998) adjusts the dendogram heights so that they form a nondecreasing sequence. The dendogram (Figure 4.11, upper left) also makes the clusters look better separated than they actually are because the within cluster variation is masked by using cluster centroids. All results reported for the Eisen clustering in this chapter are based on complete data. In fact, use of imputed data in the Eisen algorithm makes its results much more like those of other bottom-up methods.

Figure 4.13 and Figure 4.14 give dendograms for bottom-up mean linkage clustering and Eisen's clustering. They agree on some data features, such as the grouping of the normal breast tissues together. Some disagreements occur, such as the samples norway.H4 and norway.43, which are put close together by Eisen, but not joined until the top node by mean linkage. Additional methods for comparing dendograms are introduced in Section 4.5.3.

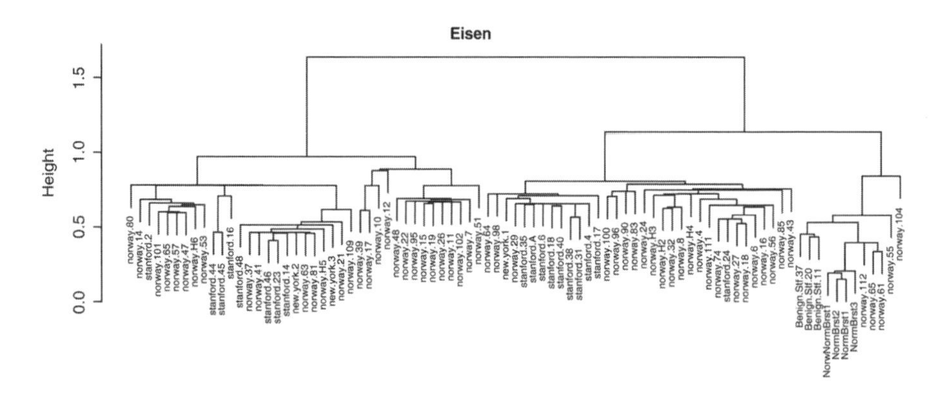

Figure 4.13 *Bottom-up Eisen clustering.*

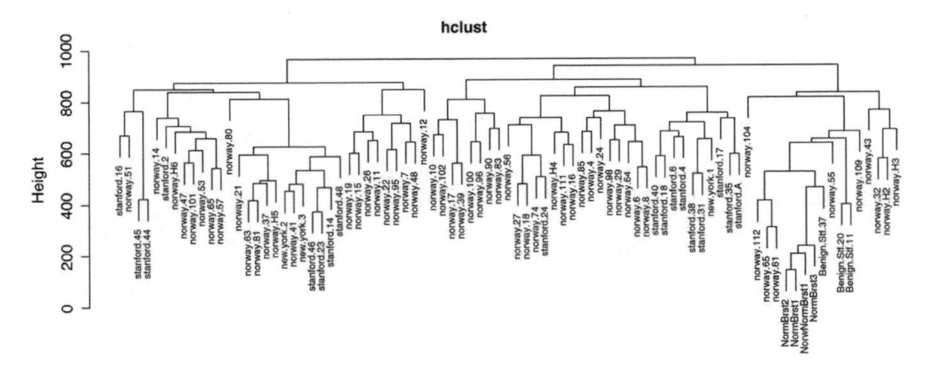

Figure 4.14 *Bottom-up mean linkage clustering.*

4.5.2 Top-down methods

To understand why top-down methods are of interest, it is useful to consider weaknesses of bottom-up methods. Bottom-up methods can poorly reflect structure near the top of the tree because many joins have been made at this stage. Each join depends on all previous joins, so if some questionable joins are made early on, they cannot be later undone. In the bottom-up method applied in Figure 4.6, observations 2 and 4 are not joined until the very top of the tree, yet the corresponding points are quite close. The partitioning presented in Figure 4.7 for $k = 2$ clusters is (arguably) superior. If interest focuses on identifying a few clusters, then top-down algorithms which successively divide observations are likely to produce more sensible partitions. Of course, these methods are less likely to give good groupings after many splits are made. This suggests that hybrids that combine the best of top-down and bottom-up methods may be useful.

There are several variations on top-down clustering, each offering a different approach to the combinatorial problem of subdividing a group of objects into two subgroups. Unlike the bottom-up case, where the best join can be identified at each step, the best partition cannot usually be found and methods attempt to find a local optimum. In the next two sections, we explore two such methods: tree structured vector quantization, which is based on K-means, and an algorithm developed by Macnaughton-Smith et al. (1965).

Tree structured vector quantization

The fact that 2-means produced a more sensible partition of the data in Figure 4.7 suggests that the 2-means algorithm might be used to recursively partition the data into clusters. That is, run 2-means on the full dataset, and then recursively run 2-means in each sub-cluster, until each point is its own cluster. This approach has been explored in the engineering literature (Gersho and Gray, 1992), and termed *tree structured vector quantization*. K-means is commonly referred to as vector quantization in this field.

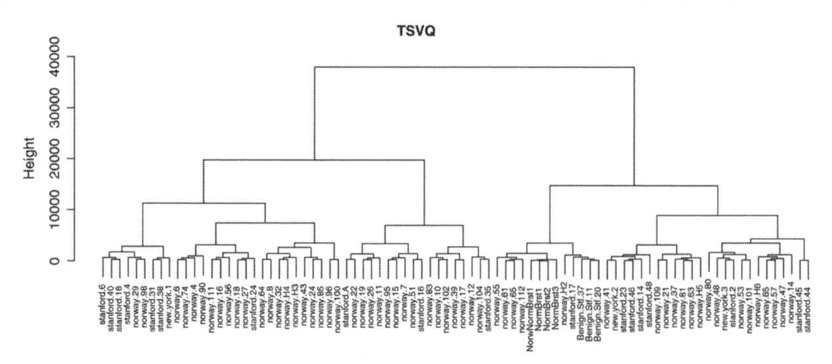

Figure 4.15 *Top-down TSVQ clustering.*

Dendograms based on this algorithm do not necessarily possess the monotonicity property of bottom-up methods. This algorithm is fast, requiring $O(n \log_2 n)$ calculations (worst-case $O(n^2)$). As with K-means, we use imputed data to deal with missing values.

Methods such as K-medoids described earlier could also be applied to this problem.

For the breast carcinoma data, a dendogram representing TSVQ is given in Figure 4.15. The most striking difference is the large distances near the top of the tree. The 2-means algorithm used to subdivide each node uses a sum of distances from the centroid instead of a mean distance. Thus, the vertical axis is a sum of distances instead of a mean distance. At early stages, the large number of points in each cluster means that improvements in subdividing are larger.

Macnaughton-Smith

Another approach to the binary partitioning problem is proposed by Macnaughton-Smith et al. (1965). In this, a "splinter group" is formed by selecting the point with greatest mean dissimilarity to all other points. Objects that are closer to the splinter group are swapped to that group one at a time until no observation in the original group is closer (in average dissimilarity) to the splinter group.

This binary partitioning is recursively carried out until each observation is separated. This top-down method produces a dendogram that obeys the monotonicity property.

Macnaughton-Smith is likely to be slower than TSVQ. Because it operates on dissimilarities instead of original data, it is easy to deal with missing values. It is also more robust to outliers because distance is used instead of squared distance.

In the breast carcinoma example, we used the Kaufman and Rousseeuw (1990) implementation of the Macnaughton-Smith algorithm, called "diana." The dendogram for the breast carcinoma data is given in Figure 4.16. The clustering is quite similar to bottom-up mean linkage clustering, although the ordering of nodes in the plot is different.

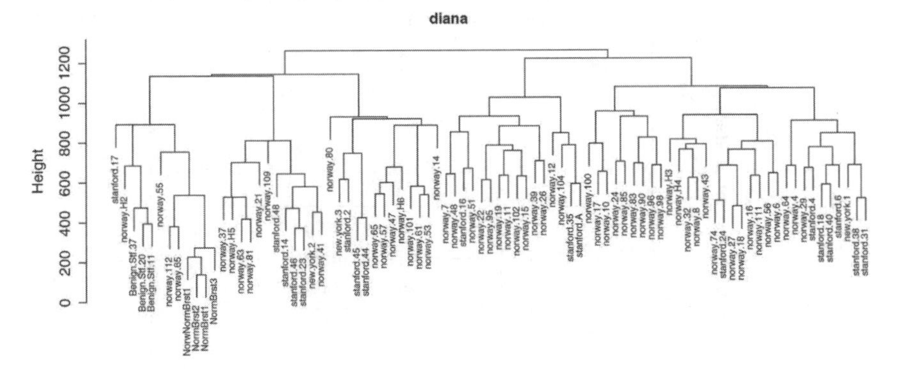

Figure 4.16 *Top-down Macnaughton-Smith clustering.*

4.5.3 Comparing the dendograms

Having described such a variety of tree-growing methods, it may be of interest to compare these trees. One such method is to look at how the trees partition up the observations into clusters for fixed cluster sizes. Suppose that we cut two trees T_1 and T_2 so each divides the data into K clusters. One measure of the dissimilarity between two trees is the proportion of all pairs of observations that are grouped together in one tree and separately in another tree. This can be calculated as follows: Let $I_1(i, i')$ be 1 if T_1 places observations i and i' in the same node and zero otherwise. One measure of similarity between trees is then

$$d(T_1, T_2) = \frac{\sum_{i>i'} |I_1(i,i') - I_2(i,i')|}{\binom{n}{2}}. \tag{4.7}$$

This was suggested in Chipman et al. (1998) in the context of classification and regression trees. The factor $\binom{n}{2}$ scales the dissimilarity to the range (0,1) with 0 indicating perfect agreement. This quantity would be calculated for all pairs of trees of interest for varying values of K. Such a comparison can be made for any clustering, and not just hierarchical ones.

In the breast carcinoma example, we compare a number of different clusterings, including the four hierarchical methods described in Section 4.5 (including several variations on bottom-up clustering), and the partitioning methods in Section 4.4. We present an informative subset of the large number of possible pairwise comparisons, starting with three variations on bottom-up clustering: single, complete, and mean linkage. Figure 4.17 shows a comparison for 2 to 20 clusters. We see that average and complete linkage result in quite similar clusterings, disagreeing on 5 to 20% of pairwise groupings. Both are quite different from single linkage clustering, which is to be expected, because single linkage can produce long chain-shaped clusters. Differences between single linkage and other methods will eventually diminish as the number of clusters approaches the sample size N. We shall not consider single

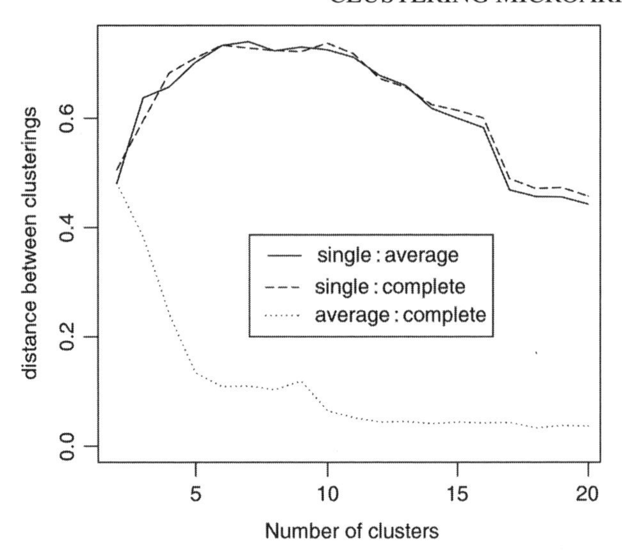

Figure 4.17 *Comparison of partitions from bottom-up hierarchical methods: single, average, and complete linkage.*

linkage further here, although the dramatic difference from other bottom-up methods suggests that in some cases it might give an alternate clustering of points.

Similar comparisons (not shown) between the Macnaughton-Smith algorithm and bottom-up mean linkage indicated very similar clusterings.

Figure 4.18 compares three hierarchical methods (Eisen, bottom-up mean linkage, TSVQ) and K-means. Although considerable disagreement exists among most methods for five or less clusters, the differences become more moderate as the number of clusters increases. The clustering produced by the Eisen algorithm is less similar to the other three. This difference between Eisen and the other three methods was more pronounced when data with missing values were used, perhaps reflecting the unusual treatment of missing values mentioned in Section 4.5.1. The similarity between K-means and TSVQ for moderate numbers of clusters is perhaps not surprising because TSVQ can be viewed as a hierarchically constrained version of K-means.

Comparisons between K-means and the PAM implementation of K-medoids (not shown) indicate large differences between two clusters. For more than five clusters, the differences are minor.

Comparisons among the hierarchical methods could be made for additional variables not used in clustering, as in Figure 4.9. In this section we omit such plots.

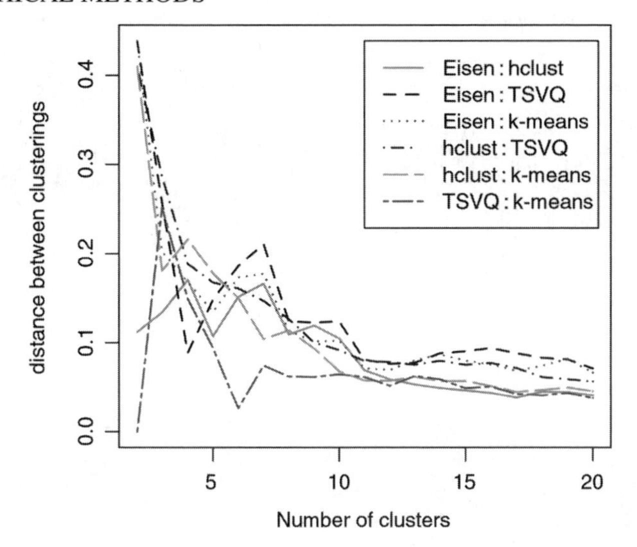

Figure 4.18 *Comparison of partitions from various methods: k-means, bottom-up (mean linkage, Eisen), and top-down (TSVQ).*

4.5.4 Post-processing of dendograms

When plotting a dendogram, the left and right children of any branch can be flipped while remaining true to the sequence of hierarchical clustering operations. In the analysis of microarray data, this reordering is key because the heat map is rearranged according to the hierarchical clustering. Visual inspection of the heatmap relies on a reasonable adjacency pattern of objects.

It is infeasible to evaluate all 2^{N-2} possible reorderings, but several heuristics have been proposed. One strategy (the default in S-Plus and R) is to order joins so that the tightest cluster (i.e., the cluster joined first) is on the left. Such an ordering is given by the dendogram in Figure 4.6. Evidently, better clusterings are possible because, with this ordering, point 2 is much closer to clusters (3,4) and (1,6) and to point 5. Sherlock's (1999) XCluster software package employs the following bottom-up reordering strategy: When two nodes are about to be joined, they are both rotated so that two most similar outermost objects in the two groups being joined are placed adjacent to each other. For example, in Figure 4.6, one reordering would be to flip the top branch so points 2 and 4 are adjacent. Here, similarity is measured in terms of only the outermost objects, ignoring interior objects. An alternative would be to calculate an average similarity between all observations in two sub-branches, instead of just using outermost objects. Further possible solutions might be to use one-dimensional multidimensional scaling or self-organizing maps (see Section 4.4.3).

The effect of reordering is illustrated in Figure 4.19. In the top two panels, trees with the default (tightest to left) and "average similarity" ordering (as mentioned previously) are plotted. The corresponding rearranged distance matrices are plotted

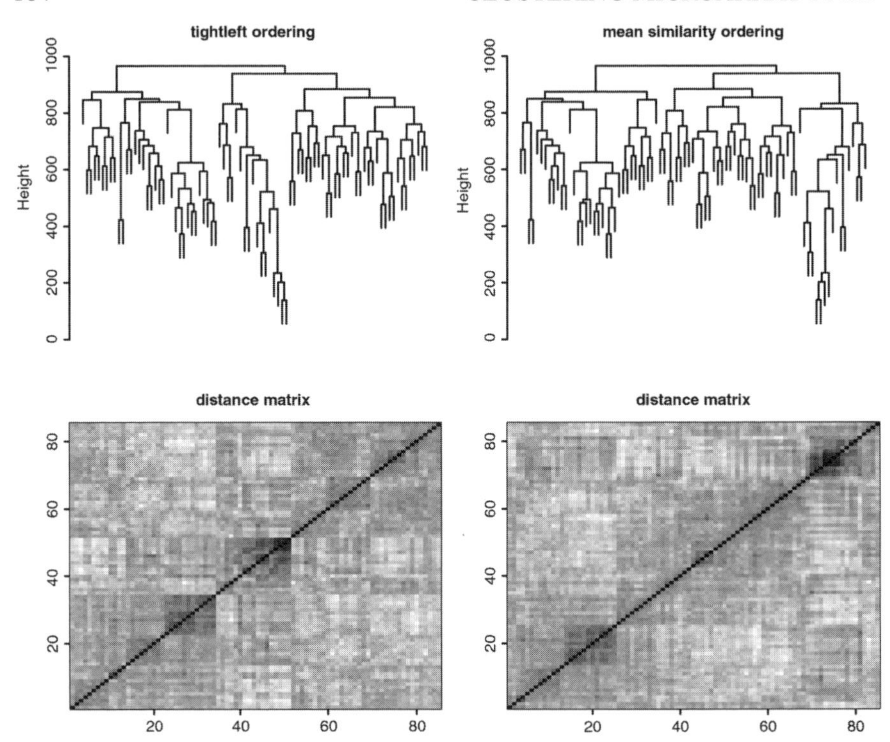

Figure 4.19 *Two orderings of the bottom-up mean linkage clustering, with corresponding dissimilarity matrices (unsquared distance is used to better highlight close clusters).*

in the bottom panels. We see that, in this case, the different arranging makes little difference. One difference is that the mean similarity ordering identifies a large group of objects in the center of the distance matrix that appear somewhat similar. Except for a few tight clusters, the overwhelming feature of this data is that the clusters found are not well separated. This can be seen in the lack of a strong block diagonal structure. This reinforces patterns seen earlier in Figure 4.8, where the decrease in sum of squares is not dramatic as the number of clusters is increased.

4.6 Two-way clustering

So far, we have discussed one-way clustering of microarray data: the samples are clustered independently of the genes, and vice versa. Typically these two operations are both used, and the rows and columns of data matrix are rearranged accordingly.

Two-way clustering methods is our name for methods that use both samples and genes simultaneously to extract joint information about both of them. For example, it may be useful to consider more than one grouping of the samples, based on different subsets of the genes. The overall clustering of the samples may be dominated by a set of genes involved say in cell proliferation, and this may mask another interesting grouping that is supported by a different set of genes.

We use the term "two-way clustering" to loosely describe methods for this problem. These methods are not yet well-developed, and are not yet in widespread use, so our discussion will be brief.

4.6.1 Coupled two-way clustering

A major difficulty with methods for clustering microarray samples is that the clustering results can depend strongly on the set of genes used. For example, use of all available genes or genes that vary at least twofold over some number of samples can produce very different hierarchical clusterings of the samples. One can either experiment with different gene lists or try to systematically undercover multiple sample clusterings based on different sets of genes. The latter approach is discussed next.

Getz et al. (2000) propose a two-way clustering method that aims to finds subsets of the genes that result in stable clusterings of a subset of the samples. That is, they find pairs $(O_k, F_k), k = 1, 2, \ldots K$ where O_k is a subset of the genes and F_k is a subset of the samples, so that when genes O_k are used to cluster samples F_k, the clustering yields stable and significant partitions. This can be especially useful when the overall clustering of the samples based on all genes is dominated by some subset of the genes, for example genes related to the profileration of the cells. By identifying and removing this first pair (O_1, F_1), other more subtle clusterings can emerge.

One could develop such an algorithm for this problem using as a building block the hierarchical cluster procedures described in this chapter. The authors in Getz et al. (2000), however, instead use an algorithm called "superparamagnetic clustering," based on a model of magnetic fields, and combine this with a complicated two-way clustering scheme. Hence, it is difficult to assess their results, and compare them to those from more standard hierarchical clustering; but their results do appear promising, and their stated goal—to find stable clustering pairs (O_k, F_k)—is one worthy of future research.

4.6.2 Block clustering

This is a top-down, row-and-column clustering of a data matrix. It reorders the rows and columns to produce a matrix with homogeneous blocks of the outcome (here, gene expression). Block clustering also produces hierarchical clustering trees for the rows and columns. The basic algorithm for forward block splitting is due to Hartigan (1972), who also reviews earlier work on two-way clustering, citing Good (1965) and Tryon and Bailey (1970). Hartigan called his approach "direct clustering," but it has become known as block clustering (e.g., Duffy and Quiroz, 1991).

Here is an outline of the block clustering procedure (see also Figure 4.20):

- Begin with the entire data in one block.
- At each stage, find the row or column split of all existing blocks into two pieces, choosing the one that produces the largest reduction in the total within block variance.
- Use only allowable splits: if there are existing row splits that intersect the block, one of these must be used for the rows, called a "fixed split." The same is done for columns. Otherwise all split points are tried.
- The splitting is continued until a large number of blocks are obtained, and then some block are recombined until the optimal number of blocks is obtained (see discussion of this point next).

To find the best split into two groups, one can show that it is sufficient to sort the rows (or columns) by row (or column) mean, and then seek a split in that order. A drawback of block clustering when applied to median centered data (which is the case here) is that at the start, all row and columns means are approximately zero. Hence, the procedure has difficulty getting started.

Restricting the choice to fixed splits ensures that:

1. The overall partition can be displayed as a contiguous representation, with a common reordering for the rows and columns.
2. The partitions of each of the rows and columns can be described by a hierarchical tree that has been cut at an appropriate level.

An alternative strategy would be to use tree-structured classification (CART; Breiman et al., 1984). Treating the rows and columns as unordered categorical variables, CART would find partitions of genes and samples that have nearly constant expression levels. However, CART does not give an overall partition that could be displayed as a contiguous representation, with a common reordering for the rows and columns, as does block clustering.

Figure 4.20 provides a simple example for illustration. It consists of 5 genes and 3 samples, labeled A–E and 1–3, respectively. The first (vertical) split separates sample 3 from 1 and 2. The second (horizontal) split separates genes B and C from A,D,E Now consider splitting the rightmost box. The split that separates genes A and B from C,D,E in the right box would not allow a single contiguous representation of the entire data matrix, and hence is not permitted. The split that separates gene B from A,C,D,E violates the hierarchical tree property (2) and is also not permitted. The only permissible horizontal split of the rightmost box is the one that separates genes B and C from A,D,E continuing the horizontal line segment in the left box all the way to the right.

We have experimented with block clustering for microarray data, and it has some potential. However, it has one major limitation: It provides only one reorganization of samples and genes, and hence cannot discover multiple groupings (as discussed previously).

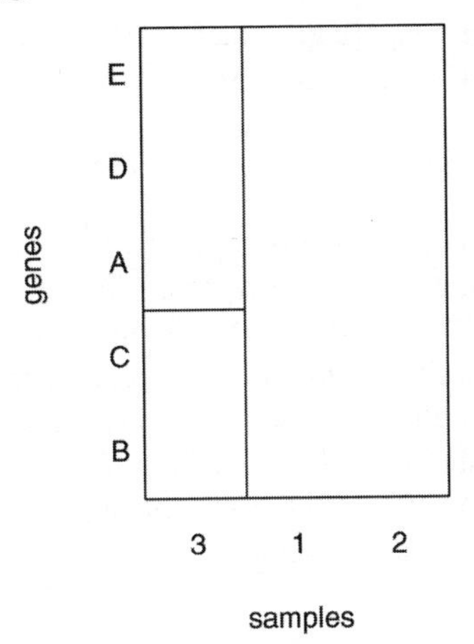

Figure 4.20 *Simple example to illustrate the block clustering rules. The first (vertical) split separates sample 3 from 1 and 2. The second (horizontal) split separates genes B and C from A,D,E. If the rightmost box is split horizontally, it must be split between genes B,C and A,D,E.*

4.6.3 Plaid models

In some sense, plaid models (Lazzeroni and Owen, 2002) generalize both block clustering and gene shaving (described in Section 4.7). Suppose Y_{ij} is the gene expression level for gene i in sample j, for $i = 1, 2, \ldots p$ and $j = 1, 2, \ldots n$. The simplest plaid model has the form

$$Y_{ij} \approx \mu_0 + \sum_{k=1}^{K} \mu_k \rho_{ik} \kappa_{jk}. \tag{4.8}$$

Here, k indexes each of the blocks or layers in the model. Each layer is rectangular, consisting of a subset of the genes and samples: $\rho_{ik} = 1$ if gene i is in layer k and zero otherwise; $\kappa_{jk} = 1$ if sample j is in layer k and zero otherwise. Within the layer, the expression level is the constant μ_k.

Note that different layers may overlap, that is, the same gene or sample may appear in more that one layer. A more general plaid model allows the expression to vary systematically within a layer:

$$Y_{ij} \approx \mu_0 + \sum_{k=1}^{K} (\mu_k + \alpha_{ik} + \beta_{jk}) \rho_{ik} \kappa_{jk}. \tag{4.9}$$

The presence of the terms α_{ik} and β_{jk} can lead to a striped pattern in the layer, and hence the name "plaid."

Estimation of the model in Equation 4.9 proceeds in a forward stagewise manner, alternatively estimating the indicator parameters ρ_{ik} and κ_{jk}, and the structural parameters $\mu_k, \alpha_{ik}, \beta_{jk}$. A permutation distribution is used to help decide when to stop adding layers (i.e., the value of K).

Figure 4.21 illustrates yeast expression data from DeRisi et al. (1997). The description of the data and plaid analyses are taken from Lazzeroni and Owen (2002). The data are available at `http://rana.stanford.edu/clustering`. The columns represent time points within each of ten experimental series.

The columns in this data correspond to ten different experiments and are denoted by the following prefixes: alpha (columns 1–18), Elu (19–32), cdc (33–47), spo (48–53), spo5 (54–56), spo- (57–58), heat (59–64), dtt (65–68), cold (69–72), diau (73–79). Experiments 1–3 examine the mitotic cell cycle. Experiments 4–6 track different strains of yeast during sporulation. Experiments 7–9 track expression following exposure to different types of shocks. Experiment 10 studies the diauxic shift. The rows are ordered according to the results of the hierarchical clustering algorithm to illustrate the relationships revealed by that approach.

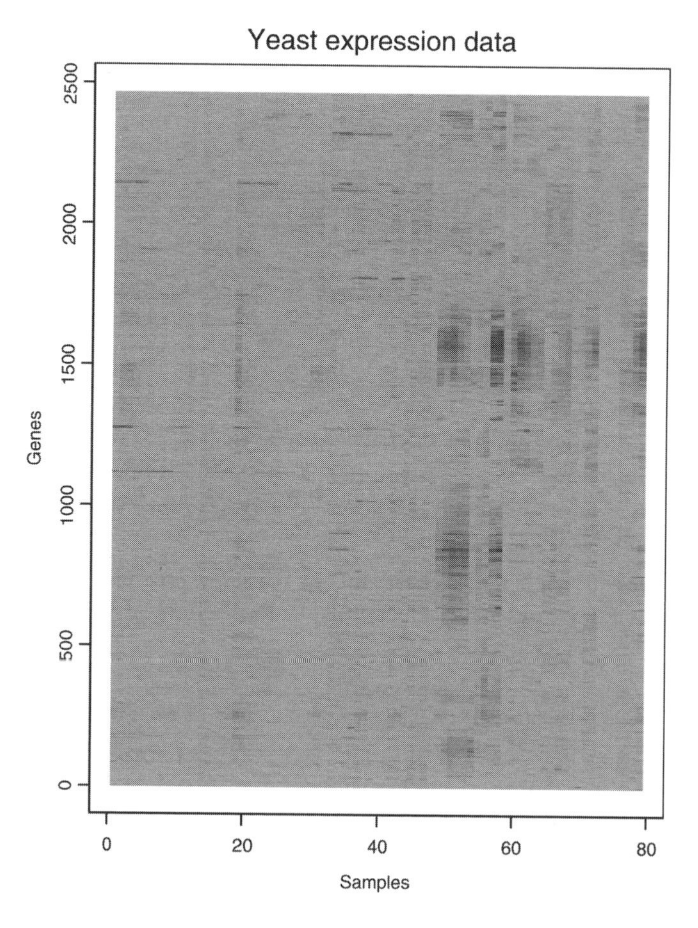

Figure 4.21 *(See color insert following page 114.) Yeast expression data.*

The authors fit a model of the form (Equation 4.9), with a total of 34 layers. They reported that no significant improvement occurred after that. The complete model is shown in Figure 4.22 and contained 5568 parameters, fewer than 3% of the number of observations. The model recovers much of the structure in the original data. Table 4.2 summarizes the gene and sample memberships.

The plaid model consistently puts columns from the same experimental series together within layers: Table 4.3 gives information about the experiments in the first 6 layers. Table 4.4 shows the genes that appear in two chosen layers, 1 and 3. These layers contain the same seven samples and share no genes in common. Layer 1 includes many genes involved in the cell cycle. Layer 3 includes many genes involved in glycolysis. In Lazzeroni and Owen (2002), the authors give details of biological insights that can be gleaned from this model.

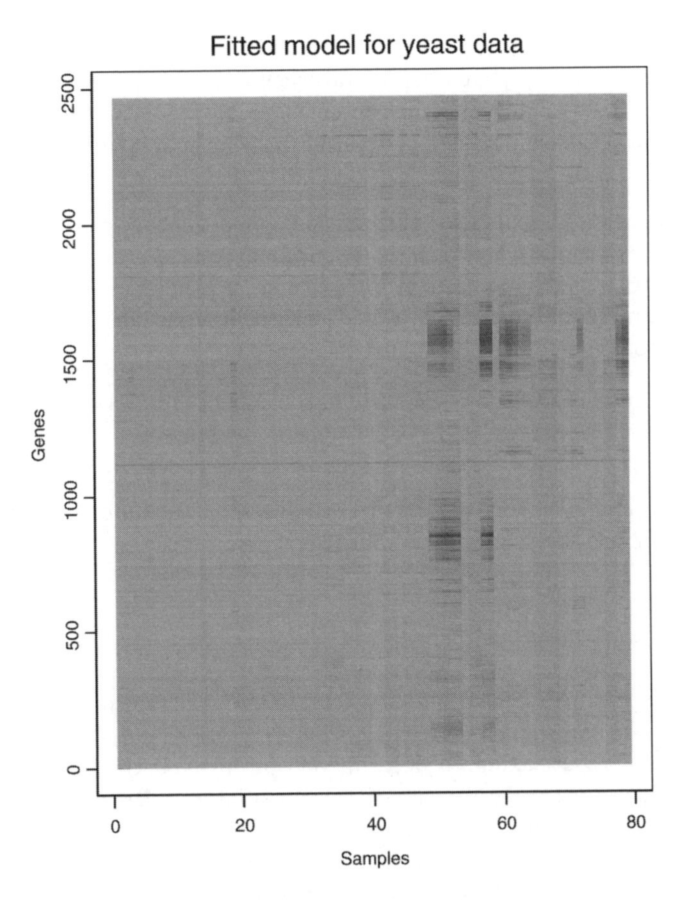

Figure 4.22 *(See color insert following page 114.) Fitted plaid model for yeast expression data.*

Table 4.2 *Yeast summary showing the numbers of genes, samples, and observations appearing in zero, one, or more layers.*

No. of Layers	Genes	Samples	Observations
0	703	22	170,703
1	1031	5	22,872
2	579	2	1307
3	142	11	11
4 – 18	12	39	0
Total	2467	79	194,893

4.7 Principal components, the SVD, and gene shaving

Principal components are derived linear combinations of a group of variables that represent them well. They are often useful when one has a large, correlated set of variables and would like to reduce them to a manageable few. Their ability to represent the original data can be measured in terms of the percent of total variation that they explain.

Some of their earliest applications of principal components where in psychometric tests and questionnaires, where many of the questions were measuring the same underlying trait, such as intelligence, but in different and correlated ways. Likewise, in gene expression arrays we have many correlated genes being co-expressed (under or over), often in response to the same biological phenomenon.

The SVD, or singular value decomposition, is an algorithm for computing the principal component analysis (PCA) of a data matrix, and the two terms are often used synonymously.

Gene shaving is an adaptation of PCA that looks for small subsets of genes that behave very similarly, and account for large variances across samples.

4.7.1 Principal Components

We describe principal components first for a set of real valued random variables $X_1, \ldots, X_p$. We seek a single, derived variable $Z_1 = a_{11}X_1 + a_{21}X_2 + \cdots + a_{p1}X_p$ such that $\text{var}(Z_1)$ is maximized. Here the a_{i1} are real-valued coefficients. If left uncontrolled, we can make the variance of Z_1 arbitrarily large. So we constrain $a_1 = (a_{i1})$ to have unit Euclidean norm, $||a_1|| = 1$. This largest principal component attempts to capture the common variation in a set of variables via a single, derived variable. Often one is not enough, so we then look for a second derived variable, Z_2, uncorrelated with the first, with the largest variance, and so on.

Table 4.3

Sample effects ($\mu + \beta_j$) in first six layers.[a]						
Sample	1	2	3	4	5	6
19 Elu 0	—	—	—	0.74	—	—
39 cdc 130	—	—	—	—	—	0.44
40 cdc 150	—	—	—	—	—	0.29
41 cdc 170	—	—	—	—	—	—
42 cdc 190	—	—	—	—	—	0.52
43 cdc 210	—	—	—	—	—	0.46
44 cdc 230	—	—	—	—	—	0.53
45 cdc 250	—	—	—	—	—	0.82
46 cdc 270	—	—	—	—	—	0.64
47 cdc 290	—	—	—	—	—	0.89
49 spo 2	0.72	–1.18	–0.81	—	—	—
50 spo 5	1.10	–1.18	–1.21	—	—	—
51 spo 7	1.36	–1.32	–1.12	—	—	0.93
52 spo 9	1.08	–0.75	–1.33	—	—	0.99
53 spo 11	1.06	—	–1.12	—	—	0.92
55 spo5 7	—	—	—	—	—	0.94
56 spo5 11	—	—	—	—	—	0.76
57 spo-early	1.19	–2.14	–1.03	—	—	—
58 spo-mid	1.41	–2.19	–1.43	—	—	—
60 heat 10	—	–1.19	—	1.57	–1.06	—
61 heat 20	—	–1.70	—	1.10	–1.15	—
62 heat 40	—	–1.23	—	0.61	–1.00	—
63 heat 80	—	–0.70	—	0.53	–0.55	—
64 heat 160	—	–0.80	—	—	–0.65	—
66 dtt 30	—	—	—	—	—	0.55
67 dtt 60	—	—	—	—	—	—
68 dtt 120	—	—	—	0.47	–0.31	—
71 cold 40	—	—	—	—	–0.59	—
72 cold 160	—	–0.90	—	—	—	—
77 diau e	—	—	—	0.55	—	—
78 diau f	—	–1.20	—	1.30	–0.64	—
79 diau g	—	–1.60	—	1.42	–0.87	0.48

[a]The rows of the table are the column effects for the first six layers of the yeast expression data. Column effects that do not appear in these layers are omitted, unless they fall between two timepoints included in a single layer.

Table 4.4 *Top* 12 *genes of layers* 1 *and* 3.

$\mu = 1.13$	Layer 1
α_i	Gene, known function
3.34	ECM11, cell wall biogenesis
2.77	LEU1, leucine biosynthesis, 3-isopropylmalate dehydratase
2.65	PDS1, cell cycle, anaphase inhibitor (putative)
2.35	CDC5, cell cycle, G2-M protein kinase
2.02	CIK1, cytoskeleton, spindle pole body associated protein
1.77	CLB5, cell cycle, G1-S cyclin
1.64	PCH2, meiosis, checkpoint
1.56	STU2, cytoskelton, spindle pole body component
1.56	BAT1, branched chain amino acid, transaminase
1.53	ORC3, DNA replication, origin recognition complex
1.56	APC4, cell cycle, anaphase-promoting complex subunit
1.51	MIP6, MRNA export, putative, RNA-binding proteinlization
$\mu = -1.20$	Layer 3
α_i	Gene, known function
–2.11	TDH1, glycolysis, glyceraldehyde-3-phosphate dehydrogenase 1
–2.02	TKL1, pentose phosphate cycle, transketolase
–1.99	PGK1, glycolysis, phosphoglycerate kinase
–1.97	ENO2, glycolysis, enolase II
–1.86	TDH2, glycolysis, glyceraldehyde-3-phosphate dehydrogenase 2
–1.79	YGP1, diauxic shift, response to nutrient limitation
–1.70	TDH3, glycolysis, glyceraldehyde-3-phosphate dehydrogenase 3
–1.68	TPI1, glycolysis, triophosphate isomerase
–1.59	FBA1, glycolysis, aldolase
–1.52	BUD7, bud site selection
–1.49	GPM1, glycolysis, phosphoglycerate mutase
–1.42	ALD6, ethanol utilization, acetaldehyde dehydrogenase

This recipe gets translated directly for a set of data, which for us will be an expression array x_{ij}, with i indexing one of the p genes (variables), and j as one of the n samples. The largest sample principal component z_{1j} is defined to be that linear combination

$$z_{1j} = \sum_{i=1}^{p} a_{i1} x_{ij}, \quad ||a_1|| = 1, \tag{4.10}$$

having largest sample variance.

Suppose that $\mathbf{S}$ is the $p \times p$ covariance matrix of the genes, with hith entry

$$S_{hi} = \frac{1}{n-1} \sum_{j=1}^{n} (x_{hj} - \bar{x}_h)(x_{ij} - \bar{x}_i), \tag{4.11}$$

where $\bar{x}_h$ and $\bar{x}_i$ are the sample means for genes h and i respectively. Then it is easy to show that the loading-vector for the first sample principal component a_1 is the largest eigenvector of $\mathbf{S}$. The second largest eigenvector a_2 defines the second principal component, and so on.

Suppose $\mathbf{A}$ is the $p \times k$ matrix consisting of the first k eigenvectors of $\mathbf{S}$; that is,

$$\mathbf{SA} = \mathbf{A\Lambda}, \tag{4.12}$$

where the diagonal matrix $\mathbf{\Lambda}$ contains the eigenvalues $\lambda_1 \geq \lambda_2 \geq \cdots \geq \lambda_k \geq 0$. Then $\mathbf{Z} = \mathbf{A}^T\mathbf{X}$ is the $k \times n$ matrix consisting of the first k principal components of the genes, also known as *eigengenes*. Figure 4.23 depicts some eigengenes for an expression array for investigating subclasses of small, round, blue-cell tumors (Khan et al., 2001; SRBCT). A total of 63 samples or arrays occur, based on either a cell-line or tissue sample, and 2308 genes. In addition the samples are classified into four subclasses labeled "EWS," "RMS," "NB," or "BL." Note that the class labels were not used in the production of the eigengenes.

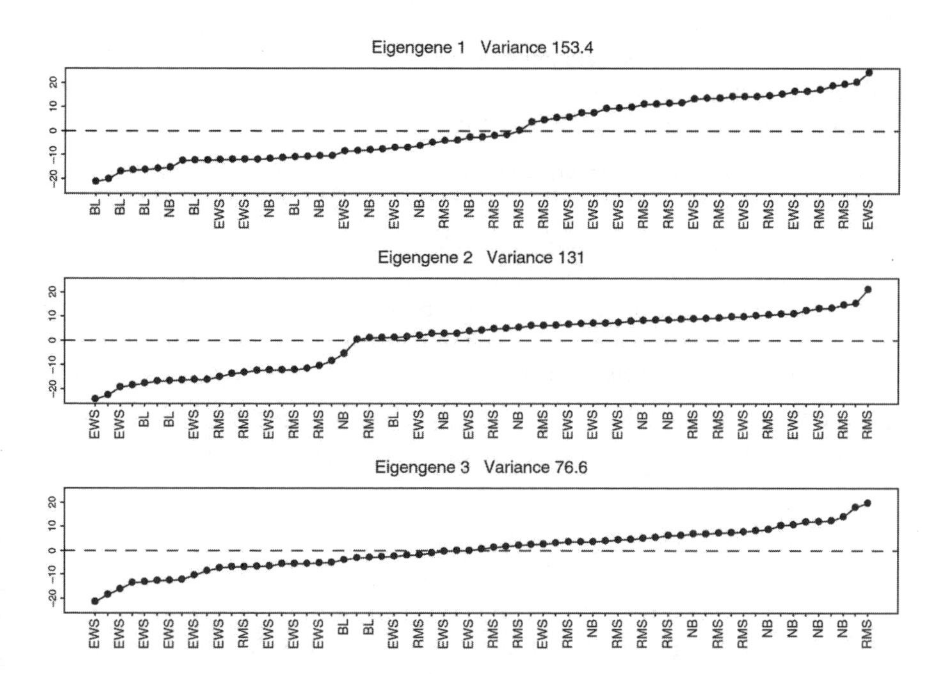

Figure 4.23 *The first three eigengenes for an expression array. Each of the samples corresponds to a cell-line or tissue sample from a small, round, blue-cell tumor, for which four subclasses are defined. The eigengenes are each sorted, and the subclass labels indicate some class grouping, especially for the first and third eigengene.*

4.7.2 Computations and the singular value decomposition

The matrix $\mathbf{S}$ is typically huge, with dimensions equal to the number of genes p, so the approach as stated is computationally unattractive. Fortunately there is a shortcut, when $n \ll p$, as is typically the case with expression arrays.

Denote by $\widetilde{\mathbf{X}}$ the expression array centered so that each row (gene) has sample mean zero. We construct its SVD

$$\widetilde{\mathbf{X}} = \mathbf{UDV}^T, \tag{4.13}$$

where $\mathbf{U}$ is $p \times n$ orthogonal, $\mathbf{V}$ is $n \times n$ orthogonal, and $\mathbf{D}$ is an $n \times n$ diagonal matrix of ordered nonnegative, singular values. Because of the centering of $\widetilde{\mathbf{X}}$, there will be, at most, $n - 1$ nonzero singular values. Now,

$$\mathbf{S} = \frac{1}{n-1}\widetilde{\mathbf{X}}\widetilde{\mathbf{X}}^T \tag{4.14}$$

$$= \frac{1}{n-1}\mathbf{UD}^2\mathbf{U}^T, \tag{4.15}$$

hence, the SVD of $\widetilde{\mathbf{X}}$ also delivers the eigen-decomposition of $\mathbf{S}$. Note that since $\widetilde{\mathbf{X}}^T\widetilde{\mathbf{X}} = \mathbf{VD}^2\mathbf{V}^T$, and $\widetilde{\mathbf{X}}\mathbf{V} = \mathbf{UD}$, we can compute the SVD through the eigen-decomposition of an $n \times n$ matrix. The columns of $\mathbf{U}$ are identical to the columns of $\mathbf{A}$ in Eq. 4.12, and $(n-1)\lambda_i = d_i^2$. From the preceding, we also note that there can be at most $n - 1$ eigengenes with positive variance, and typically the first six to ten are the most useful. The columns of $\mathbf{U}$ are sometimes called *eigenarrays* (Alter et al., 2000), and the entries give an indication of which genes play a leading role in the construction of the corresponding eigengene; see Figure 4.24.

A reduced-rank SVD can be interpreted as a sparse reconstruction of the (centered) expression array

$$\widetilde{\mathbf{X}} \approx \mathbf{U}_q\mathbf{D}_q\mathbf{V}_q^T; \tag{4.16}$$

here the subscript indicates that we have used only the first q columns of $\mathbf{U}$, $\mathbf{D}$, and $\mathbf{V}$. The idea is that most of the action in the expression array can be approximated by the activity of the first q eigengenes. Each of the genes is then approximated by its regression onto this set of eigengenes.

Figure 4.25 depicts the variances of the eigengenes for the SRBCT cancer data; there is evidence that about the first ten eigengenes play a leading role in explaining the variation in the data.

4.7.3 Gene Shaving

Each eigenarray in Figure 4.24 involves all the genes, each with its own weight. This is not too helpful in forming informative subsets of the genes. Because genes tend to be activated in pathways, we expect to find groups of genes co-expressing in response to the same biological phenomenon. Gene-shaving specifically looks for subsets of genes that are strongly correlated and where the average has a large variance. These

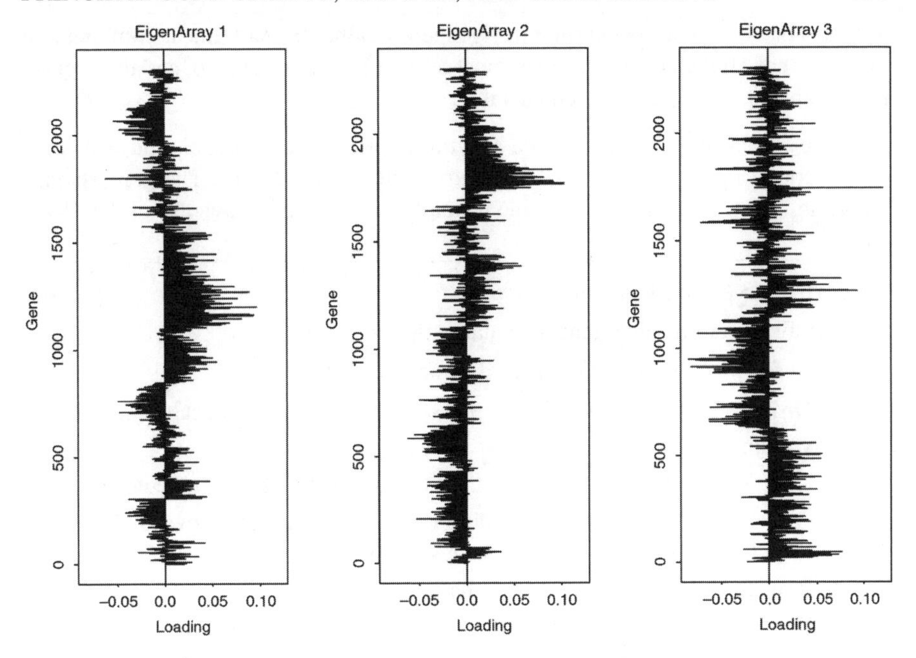

Figure 4.24 *The first three eigenarrays for the cancer data. The genes are in the order produced by a hierarchical clustering. This explains the patterns in the eigenarrays and aids in their interpretation.*

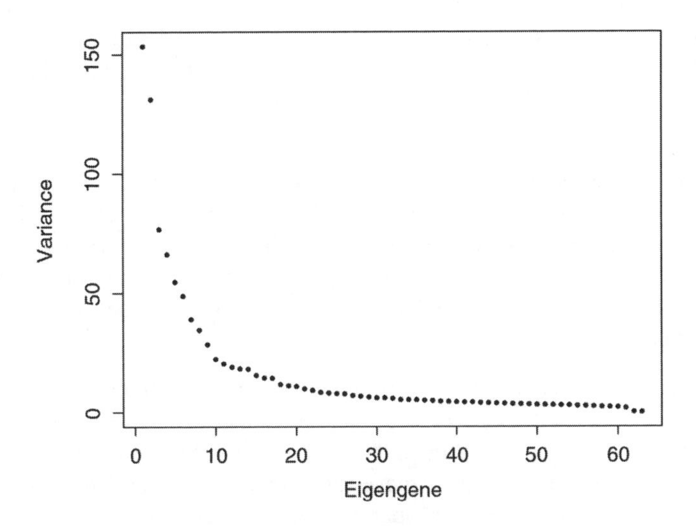

Figure 4.25 *Variances of the eigengenes. The kink in the curve suggests that about the first ten eigengenes might explain the important activity in the data.*

averages or *supergenes* are similar to eigengenes in that they are linear combinations (averages) of all the genes; however, most of their loadings are zero, and those genes in the subset have loadings of equal value.

Next, we give an idea of how the gene-shaving algorithm works; details and extensions can be found in Hastie et al. (2000a) and Hastie et al. (2000b). The gene-shaving algorithm given in the following table is iterative and makes heavy use of the SVD.

Algorithm 3 *Gene shaving algorithm.*

1. Initialize, setting supergene s_0 equal to the n-vector of ones.
2. Loop for $j = 1$ to J, the number of desired supergenes.
   a. Orthogonalize the rows of the expression array $\mathbf{X}$ with respect to $s_0, \ldots, s_{j-1}$, yielding $\widetilde{\mathbf{X}}$.
   b. Compute the SVD of $\widetilde{\mathbf{X}}$, and extract the largest left, singular vector $\mathbf{u}$.
   c. Order the elements of $\mathbf{u}$ in absolute value, and remove the rows of $\widetilde{\mathbf{X}}$ corresponding to the fraction α of smallest such values.
   d. Repeat steps 2b. and 2c. until $\widetilde{\mathbf{X}}$ has only one row.
   e. From the sequence of submatrices, select the one, as well as its *signed* row average s_j, using the *gap test* (see the following paragraphs).

Genes can remain in a subset if their SVD loadings in $\mathbf{u}$ are large in absolute value. We hence assign a sign to each gene in the subset, and apply the sign before computing an average.

The gap test for gene shaving compares the variances for each subset in the shaving sequence to a similar sequence obtained from randomized data. Specifically, each row of the expression array is randomly permuted, and the gene shaving algorithm is applied. Figure 4.26 illustrates the idea; the gap test is described in more detail in Hastie et al. (2000b).

The orthogonalization in step 2a. is implicitly performed in principal components as well, and guarantees that our supergenes will be different from each other. Figure 4.27 shows the first three shaves produced by this algorithm. Above each plot is the variance of the cluster average or supergene. Also reported is the fraction of the total variance of the genes in the cluster explained by the cluster average; a value close to 1 indicates a very tight cluster.

These data are accompanied by an additional label for each sample, classifying it into one of four cancer subclasses. Figure 4.28 shows that supergenes 1 and 3 separate the subclass labels well, even though they were not used to produce the shaves. The eigengenes of Figure 4.23 do not achieve comparable separation. The second shave is not associated with the class labels, but is a very tight cluster with an abrupt change from red to green. We examined other attributes that accompany these data, but were unable to find an explanation for the behavior of this group of genes.

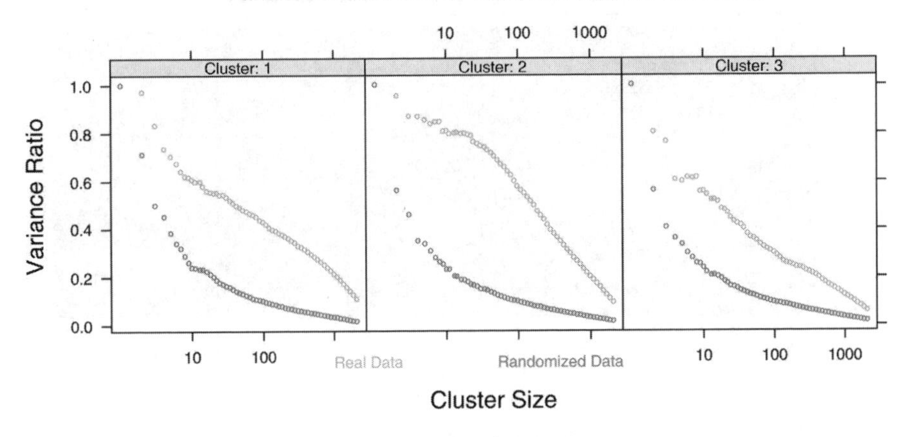

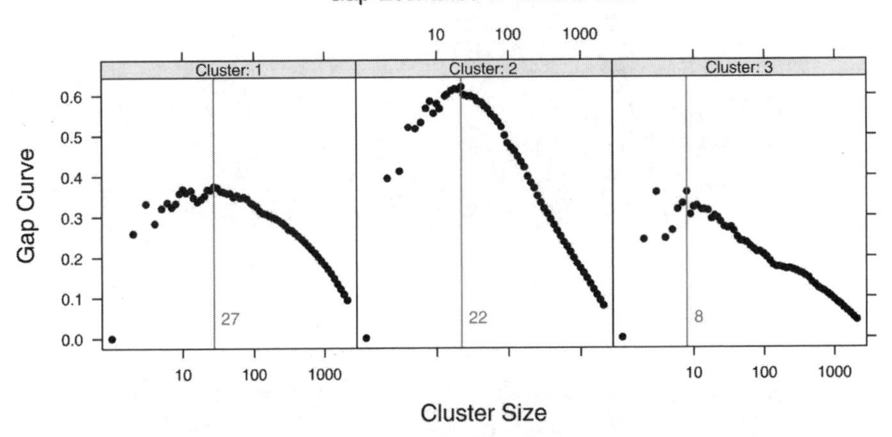

Figure 4.26 *The gap test is used to select the subset sizes in gene shaving. For each sequence of shaves, we compare the variance to that obtained by shaving a matrix with randomly permuted rows (top panels). In the lower panels, the differences or "gaps" between these curves indicate the point of largest discrepancy.*

We have simply numbered the genes in each shave here; in practice, we have available the most current information on each gene, and since the shaves are of a manageable size, the individual genes can be examined more closely and compared.

4.8 Other approaches

A great deal of research has been conducted in clustering, particularly as it is applies to microarrays. This work originates from a wide variety of fields, including statistics, biology, engineering, computer science, and physics. We have focused on some of the

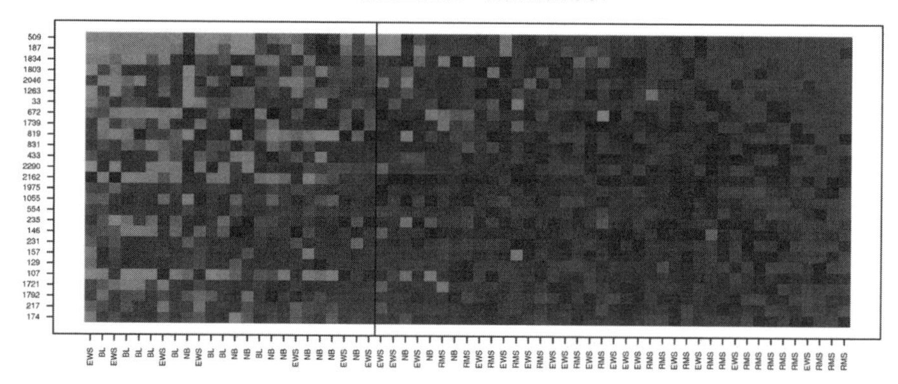

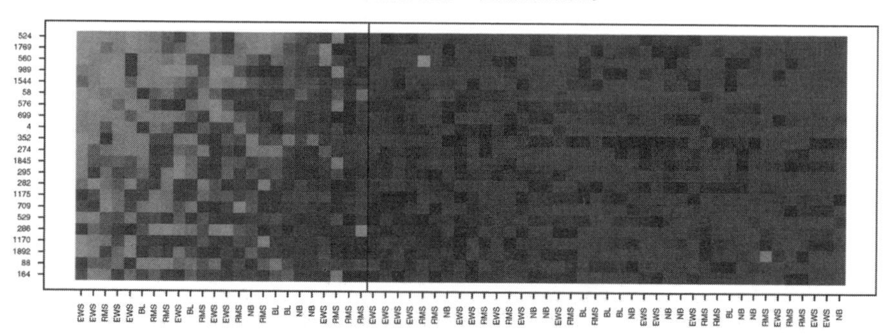

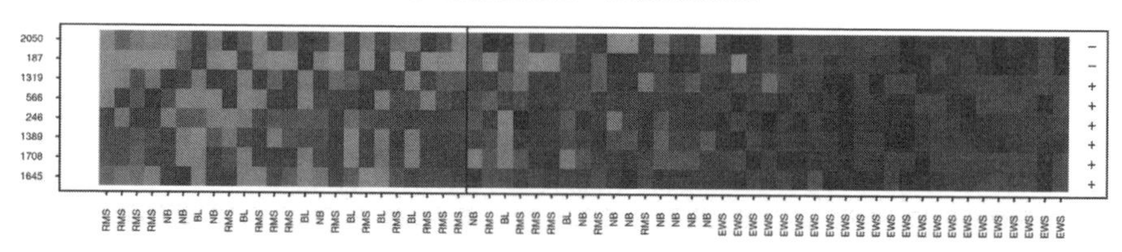

Figure 4.27 *(See color insert following page 114.) The first three shaves for the SRBCT cancer data. In each panel, we see the cluster of genes ordered according to their supergene (their signed average); columns (samples) are ordered by mean expression within each shave; the cancer subclass labels are indicated below each panel. The "+" and "−" signs in the third shave indicate that some genes had their signs flipped in the averaging.*

more widely used methods, especially on techniques that have been proposed in the statistical literature. It would be impossible to discuss most or all of the methods that have been proposed, and it is still an evolving area. The following is a list of some recent articles; they contain references to further work:

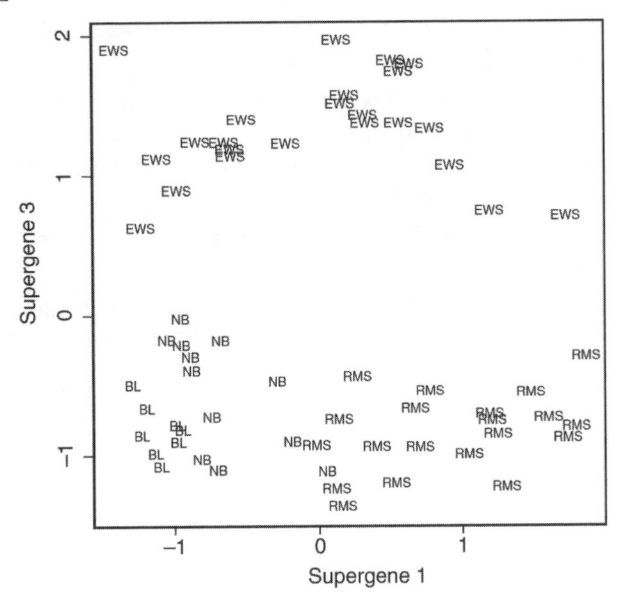

Figure 4.28 *The samples plotted in the two-dimensional subspace defined by Supergenes 1 and 3. These two derived subgroups separate the training data perfectly.*

- *Mixture models:* Heydebreck et al. (2002), McLachlan et al. (2002), and Pan et al. (2002)
- *Graph-theoretic:* Sahran and Shamir (2000)
- *Two-way:* Tang et al. (2001)
- *Comparative study:* Goldstein et al. (2002)

4.9 Software

In practice, the choice of clustering algorithm depends to a large extent on available software. Most computations in this chapter were done in R and S-Plus, two implementations of the S programming language. All R packages are available at `http://www.r-project.org/`. Specific algorithms include:

`cluster` package, S original by Struyf et al. (1997), ported to R by Kurt Hornik and Martin Maechler. This package includes K-medoids (`pam`), Macnaughton-Smith algorithm (`diana`), and a distance function able to deal with missing values (`daisy`).

`hclust` function, part of the S-Plus package, R code part of the `mva` package. The `hclust` function performs bottom-up hierarchical clustering with single, complete, mean, and other linkages. The "hclust" function in R is based on Fortran code contributed to STATLIB by F. Murtagh.

k-means function, part of the S-Plus package, R code part of the mva package. This is the Hartigan and Wong (1979) algorithm.

SOMs: Tamayo et al.'s (1999) GENECLUSTER software is available in 1.0 and 2.0 versions; see http://www-genome.wi.mit.edu/cancer/software/software.html for details. An R library GeneSOM, written by Jun Yan, is available from the R site.

Plaid models: Software is available at http://www-stat.stanford.edu/~owen/plaid/.

Custom code was developed in S-Plus and R for TSVQ, Eisen's bottom-up clustering, heatmaps, and the imputation of missing values. This code is available from the authors on request.

Michael Eisen's bottom-up clustering is available as a Windows software package (Cluster), along with Treeview, a microarray visualization package. See Eisen (1998) and http://rana.lbl.gov/EisenSoftware.htm.

References

Aach, J. and G.M. Church, Aligning gene expression time series with the warping algorithms, *Bioinformatics,* 17:495–508, 2001.

Affymetrix Inc., Array Design for the GeneChip Human Genome U133 Set, Technical Report, 2002.

Affymetrix Inc., GeneChip Expression Analysis Technical Manual, Technical Report, 2001a.

Affymetrix Inc., New Statistical Algorithms for Monitoring Gene Expression on GeneChip Probe Arrays, Technical note, http://www.affymetrix.com/pdf/algorithms.pdf, 2001b.

Alizadeh, A.A. et al., Different types of diffuse large B-cell lymphoma identified by gene expression profiling, *Nature,* 403:503–511, 2000.

Alon, U. et al., Broad patterns of gene expression revealed by clustering analysis of tumor and normal colon tissues probed by oligonucleotide arrays, *Proc. Natl. Acad. Sci. USA,* 96: 6745–6750, 1999.

Alter, O., P. O. Brown, and D. Botstein, Singular value decomposition for genome-wide expression data processing and modeling, *Proc. Natl. Acad. Sci.,* 97(18):10,101–10,106, 2000.

Ambroise, C. and G. J. McLachlan, Selection bias in gene extraction on the basis of microarray gene-expression data, *Proc. Natl. Acad. Sci.,* 99(10):6562–6566, 2002.

Baggerly, K.A. et al., Identifying differentially expressed genes in cDNA microarray experiments, *J. Computational Biol.,* 8(6):639–659, 2001.

Baldi, P. and A.D. Long, A Bayesian framework for the analysis of microarray expression data: regularized *t*-test and statistical inferences of gene changes, *Bioinformatics,* 509–519, 2001.

Bar-Joseph, Z. et al., A new approach to analyzing gene expression time series data, in *RECOMB 2002: Proc. Sixth Annu. Int. Conf. on Computational Biol.,* 2002.

Barnard, M., The secular variations of skull characters in four series of Egyptian skulls, *Annals of Eugenics,* 6:352–371, 1935.

Bedalov, A. et al., Identification of a small molecule inhibitor of sir2p. *Proc. Natl. Acad. Sci.,* 98:15,113–15,118, December 2001.

Beer, D.G. et al., Gene-expression profiles predict survival of patients with lung adenocarcinoma, *Nature Medicine,* 8(8):816–824, 2002.

Benjamini, V. and V. Hochberg, Controlling the false discovery rate: a practical and powerful approach to multiple testing, *J. R. Statist. Soc. B.,* 57:289–300, 1995.

Benjamini, V. and D. Yekutieli, The control of the false discovery rate in multiple testing under dependency, *Annals of Statistics,* 29(4):1165–1188, 2001.

Bennett, J.H., Ed., *Fisher, Ronald Aylmer, Sir, 1890–1962: Collected papers of R. A. Fisher,* University of Adelaide, 1971–1974.

Bhattacharjee, A. et al., Classification of human lung carcinomas by mRNA expression profiling reveals distinct adenocarcinoma sub-classes, *Proc. Natl. Acad. Sci.,* 98(24):13790–13795, 2001.

Bittner, M. et al., Molecular classification of cutaneous malignant melanoma by gene expression profiling. *Nature* 406:536–540, 2000.

Bowtell, D. and J. Sambrook, Eds., *DNA Microarrays: A Molecular Cloning Manual,* CSHL Press, September 2002.

Box, G.E.P., J.S. Hunter, and W.G. Hunter, *Statistics for Experimenters: An Introduction to Design, Data Analysis, and Model Building.* New York: John Wiley & Sons, 1978.

Boldrick, J.C. et al., Stereotyped and specific gene expression programs in human innate immune responses to bacteria, *Proc. Natl. Acad. Sci.,* 99(2):972–977, 2002.

Bø, T.H. and I. Jonassen, New feature subset selection procedures for classification of expression profiles, *Genome Biol.,* 3(4):1–11, 2002.

Breiman, L., Statistical modeling: the two cultures, *Statistical Science,* 16(3):199–215, 2001.

Breiman, L., Random forests-random features. Technical Report 567, Department of Statistics, University of California, Berkeley, 1999.

Breiman, L., Arcing classifiers, *Annals of Statistics,* 26:801–824, 1998a.

Breiman, L., Using convex pseudo-data to increase prediction accuracy, Technical Report 513, Department of Statistics, University of California, Berkeley, March 1998b.

Breiman, L., Bagging predictors. *Machine Learning,* 24:123–140, 1996a.

Breiman, L., Heuristics of instability and stabilization in model selection, *Annals of Statistics,* 24(6):2350–2383, 1996b.

Breiman, L., Out-of-bag estimation, Technical Report, Department of Statistics, University of California, Berkeley, 1996c.

Breiman, L. et al., *Classification and regression trees,* The Wadsworth statistics/probability series. Wadsworth International Group, 1984.

Brown, P.O. and D. Botstein, Exploring the new world of the genome with DNA microarrays, *Nature Genetics,* 21(Suppl. 1):33–37, 1999.

Brown, T.A. and J. Koplowitz, The weighted nearest neighbor rule for class dependent sample sizes, *IEEE Trans. Information Theory,* 25(5):617–619,1979.

Buhler, J. Dapple, http://www.cs.wustl.edu/rvjbuhler/research(dapple/index.html, 2002).

Buhler, J. et al., Dapple: improved techniques for finding spots on DNA microarrays. UWCSE Tech Report UWTR 2000-08-05, Department of Computer Science and Engineering, University of Washington, Seattle, WA, August 2000.

Burges, C.J.C., A tutorial on support vector machines for pattern recognition, *Data Mining and Knowledge Discovery,* 2:121–167, December 1998.

Burton, G.R. et al., Microarray analysis of gene expression during early adipocyte differentiation, *Gene,* 293:21–31, 2002.

Bussemaker, H.J., H. Li, and E. D. Siggia, Building a dictionary for genomes: identification of presumptive regulatory sites by statistical analysis, *Proc. Natl. Acad. Sci.,* 97:10096–10100, 2000.

Bussemaker, H.J., H. Li, and E. D. Siggia, Regulatory element detection using correlation with expression. *Nature Genetics,* 167–171, 2001.

Butte, A.J. et al., Comparing the similarity of time-series gene expression using signal processing metrics. *J. Biomedical Informatics,* 34:396–405, 2002.

Buttrey, S.E. and C. Karo, Using k-nearest neighbor classification in the leaves of a tree, Technical Report, Naval Postgraduate School, Monterey, CA, 2002.

Callow, M. J. et al., Microarray expression profiling identifies genes with altered expression in HDL deficient mice, *Genome Res.,* 10(12):2022–2029, 2000.

Chang, C.-C. and C.-J. Lin, Libsvm: a library for support vector machines (version 2.31), Technical Report, http://www.csie.ntu.edu.tw/ cjlin/papers/libsvm2.ps.gz, 2001.

Chen, X. et al., Gene expression patterns in human liver cancers, *Molecular Biology of the Cell,* 13(6):1929–1939,2002.

Chen, Y., E. R. Dougherty, and M. L. Bittner, Ratio-based decisions and the quantitative analysis of cDNA microarray images, *J. Biomedical Optics,* 2:364–374, 1997.

Chiang, D.Y., P. O. Brown, and M. B. Eisen, Visualizing association between genome sequences and gene expression data using genome-mean expression profiles, *Bioinformatics,* 81:849–855, 2001.

Chipman, H., E. George, and R. McCulloch, Making sense of a forest of trees, in *Proc. 30th Symposium on the Interface,* S. Weisberg, Ed., Interface Foundation of North America, Fairfax Station, VA, 84–92, 1998.

Cho, R.J. et al., A genome-wide transcriptional analysis of the mitotic cell cycle, *Molecular Cell,* 2:65–73, July 1998.

Chow, M.L., E.J. Moler, and I.S. Mian, Identifying marker genes in transcription profiling data using a mixture of feature relevance experts, *Physiological Genomics,* 5:99–111, 2001.

Chu, S. et al., The transcriptional program of sporulation in budding yeast, *Science,* 282:699–705, 1998.

Cohen, B.A. et al., A computational analysis of whole genome expression data reveals chromosomal domains of gene expression, *Nature Genetics,* 26(2):183–186, 2000.

Cortes, C. and V. Vapnik, Support-vector networks, *Machine Learning,* 20:273–297, December 1995.

Costello, J.F. et al., Aberrant CpG-island methylation has non-random and tumour-type-specific patterns, *Nature Genetics,* 24:132–138, 2000.

Cox, D.R., *Planning of Experiments,* New York: John Wiley & Sons, Wiley publications in statistics, 1958.

Cox, D.R. and D.V. Hinkley, *Theoretical Statistics.* London: Chapman & Hall, 1974.

DeRisi, J.L., V.R. Iyer, and P.O. Brown, Exploring the metabolic and genetic control of gene expression on a genomic scale, *Science,* 278:680–685, 1997.

DeRisi, J. et al., Use of a cDNA microarray to analyse gene expression patterns in human cancer, *Nature Genetics,* 14:457–460, 1996.

Dettling, M. and P. Bülmann, How to Use Boosting for Tumor Classification with Gene Expression Data, Res. Rep. 103, Sem. f. Statistik, ETH, Zurich, 2002.

Diaz, E. et al., Molecular analysis of gene expression in the developing pontocerebellar projection system, *Neuron* 36(3):417–434, 2002a.

Diaz, E. et al., Molecular analysis of positional identity in the developing mouse retina, Submitted, 2002b.

Doniger, S.W. et al., Mappfinder: using gene ontology and genmapp to create a global gene expression profile from microarray data, *J. Genome Biol.,* 4(1):R7, 2003.

Dougherty, E.R., S. Kim, and Y. Chen, Coefficient of determination in nonlinear signal processing. *Signal Process.,* 80:2219–2235, 2000.

Dudoit, S., J.P. Shaffer, and J.C. Boldrick, Multiple hypothesis testing in microarray experiments, Division of Biostatistics, University of California at Berkeley, Technical Report 110, 2002a.

Dudoit, S. et al., Statistical methods for identifying differentially expressed genes in replicated cDNA microarray experiments, *Statistica Sinica,* 12(1): 111–139, 2002b.

Dudoit, S., J. Fridlyand, and T.P. Speed, Comparison of discrimination methods for the classification of tumors using gene expression data, *J. Am. Statistical Assoc.,* 97(457):77–87,2002a.

Dudoit, S. and Y.H. Yang, Bioconductor R packages for exploratory analysis and normalization of cDNA microarray data, in *The Analysis of Gene Expression Data: Methods and Software,* Parmigiani, G. et al., Eds., New York: Springer-Verlag, 2003.

Duffy, D. and A. Quiroz, A permutation-based algorithm for block clustering, *J. Classification,* 8:65–91, 1991.

Durbin, B.P. et al., A variance-stabilizing transformation for gene-expression microarray data, *Bioinformatics,* 18:S105–S110, 2002.

Efron, B. et al., Empirical bayes analysis of a microarray experiment, *J. Am. Statistical Assoc.,* 96:1151–1160, 2001.

Efron, B. et al., Microarrays and their use in a comparative experiment, Technical Report, Department of Statistics, Stanford University, Stanford, CA, 2000.

Efron, B. and R. Tibshirani, *An Introduction to the Bootstrap,* New York: Chapman & Hall/CRC Press, 1993.

Efron, B. and R. Tibshirani, Empirical Bayes methods and false discovery rates for microarrays, *Genet. Epidemiol.,* 23(1): 70–86, 2000.

Eisen, M., ScanAlyze. http://rana.1bl.gov/EisenSoftware.htm, 1999.

Eisen, M., *Cluster and Tree View Manual,* Stanford, CA: Stanford University. Available online at http://rana.lbl.gov/EisenSoftware.html,1998.

Eisen, M. et al., Cluster analysis and display of genome-wide expression patterns, *Proc. Natl. Acad. Sci.,* 95:14863–14868, 1998.

Filkov, V., S. Skiena, and J. Zhi. Analysis techniques for microarray time-series data, *J. Computational Biol.,* 317–330, 2002.

Fisher, R.A., The use of multiple measurements in taxonomic problems, *Annals of Eugenics,* 7:179–188, 1936.

Fisher, R.A., The arrangement of field experiments. *J. Min. Agric. Gr. Br.,* 33:505–513, 1926.

Fix, E. and J. Hodges, Discriminatory analysis, nonparametric discrimination: consistency properties, Technical Report, Randolph Field, Texas: U.S. Air Force School of Aviation Medicine, 1951.

Fleury, G. et al., Clustering gene expression signals from retinal microarray data, in *Proc. IEEE Int. Conf. on Acoust., Speech, and Sig. Process.,* 2002a.

Fleury, G. et al., Pareto analysis for gene filtering in microarray experiments, in *European Signal Process. Conf.,* http//:www.eecs.umich.edu/hero/preprints/eusipcogene.pdf, September 2002b.

Freund, Y. and R.E. Schapire, A decision-theoretic generalization of on-line learning and an application to boosting, *J. Computer Syst. Sci.,* 55:119–139,1997.

Friddle, C.J. et al., Expression profiling reveals distinct sets of genes altered during induction and regression of cardiac hypertrophy, *Proc. Natl. Acad. Sci.,* 97:6745–6750, 2000.

Friedman, J., T. Hastie, and R. Tibshirani, Additive logistic regression: a statistical view of boosting (with discussion), *Annals Statistics,* 28:337–404, 2000.

Friedman, N. et al. Using Bayesian networks to analyze expression, in *RECOMB, Proc. 4th Annu. Int. Conf. on Computational Biol.,*127–135, 2000.

Friedman, J.H., Another approach to polychotomous classification,Technical Report, Department of Statistics, Stanford University, Stanford, CA, 1996a.

Friedman, J.H., On bias, variance, O/l-loss, and the curse-of-dimensionality, Technical Report, Department of Statistics, Stanford University, Stanford, CA, 1996b.

Friedman, J.H., Flexible metric nearest neighbor classification, Department of Statistics, Technical Report, Stanford University, Stanford, CA,1994.

Galitski, T. et al., Ploidy regulation of gene expression, *Science,* 285:251–254, 1999.

Gersho, A. and R. Gray, *Vector Quantization and Signal Compression,* Boston: Kluwer Academic Publishers, 1992.

Getz, G., E. Levine, and E. Domany, Coupled two-way clustering analysis of gene microarray data, *Proc. Nat. Acad. Sci.,* 12079–12084, 2000.

Ghosh, D., Resampling methods for variance estimation of singular value decomposition analyses from microarray experiments, *Functional Integrative Genomics,* 2:92–97, 2002.

Gilks, W.R., S. Richardson, and D.J. Speigelhalter, *Markov Chain Monte Carlo in Practice,* New York: Chapman & Hall, 1995.

Glonek, G.F.V. and P.J. Solomon, Factorial designs for microarray experiments, University of Adelaide, http://www.maths.adelaide.edu.au/MAG, February 2002.

Goldstein, D., D. Ghosh, and E. Conlon, Statistical issues in the clustering of gene expression data, *Statistica Sinica,* 12(1):241–262, 2002.

Golub, T.R. et al., Molecular classification of cancer: class discovery and class prediction by gene expression monitoring, *Science,* 286:531–537,1999.

Good, I., *Categorization of Classification Mathematics and Computer Science in Biology and Medicine,* London: Her Majesty's Stationery Office, 1965.

Hakak, Y. et al., Genome-wide expression analysis reveals dysregulation of myelination-related genes in chronic schizophrenia, *Proc. Natl. Acad. Sci.,* 98:4746–4751, 2001.

Hampel, F.R. et al., *Robust Statistics,* New York: John Wiley & Sons, 1986.

Han, E. et al., Pooling samples in short oligonucleotide microarray experiments, Paper in preparation, 2002.

Hartigan, J.A., Direct clustering of a data matrix, *J. Amer. Statis. Assoc.,* 6:123–129, 1972.

Hartigan, J.A. and M.A.Wong, [Algorithm AS 136] A k-means clustering algorithm (AS R39: 81v30 p355–356), *Applied Statistics,* 28:100–108, 1979.

Hastie T. and R. Tibshirani. Classification by pairwise coupling. Technical Report, Department of Statistics, Stanford University, 1996a.

Hastie T. and R. Tibshirani, Discriminant adaptive nearest-neighbor classification, *IEEE Pattern Recognition and Machine Intelligence,* 18:607–616, 1996b.

Hastie, T., R. Tibshirani, and J. H. Friedman, *The Elements of Statistical Learning: Data Mining, Inference, and Prediction,* New York: Springer-Verlag, 2001.

Hastie, T. et al., Identifying distinct sets of genes with similar expression patterns via gene shaving, *Genome Biol.,* 1(2):1–21, 2000a.

Hastie, T. et al., Gene shaving: a new class of clustering methods for expression arrays, Technical Report, Department of Statistics, Stanford University, Stanford, CA, 2000b.

Hastie, T. et al., Imputing missing data for gene expression arrays, Technical Report, Department of Statistics, Stanford University, Stanford, CA, 1999.

Hero, A. and G. Fleury, Pareto-optimal methods for gene filtering, http://www.eecs.umich.edu/hero/bioinfo.html, 2002.

Heydebreck, A. et al., Identifying splits with clear separation: a new class discovery method for gene expression data, *Bioinformatics,* 107–114, 2002.

Hofmann, W.K. et al., Relation between resistance of Philadelphia-chromosome-positive acute lymphoblastic leukaemia to the tyrosine kinase inhibitor STI571 and gene-expression profiles: a gene-expression study, *Lancet* 359(9305):481–486, 2002.

Holder, D. et al., Statistical analysis of high-density oligonucleotide arrays: a SAFER approach. *Proc. Am. Statistical Assoc.,* Atlanta, GA, 2001.

Holter, N. et al., Dynamic modeling of gene expression data, *Proc. Natl. Acad. Sci.,* 98(4):1693–1698, 2001.

Huber, P.J., *Robust Statistics,* New York: John Wiley & Sons, 1981.

Huber, W. et al.,Variance stabilization applied to microarray data calibration and to the quantification of differential expression, *Bioinformatics,*18:896–8104, 2002.

Hughes, J.D. et al., Computational identification of cis-regulatory elements associated with functionally coherent groups of genes in saccharomyces cerevisiae, *J. Molecular Biol.,* 296:1205–1214, 2000a.

Hughes, T.R. et al., Expression profiling using microarrays fabricated by an ink-jet oligonucleotide synthesizer. *Nature Biotechnol.,*19:342–347, 2001.

Hughes, T.R. et al., Functional discovery via a compendium of expression profiles, *Cell,*102:109–126, 2000b.

Ideker, T. et al., Testing for differentially expressed genes by maximum-likelihood analysis of microarray data, *J. Computational Biol.,*7(6):805–817, 2000.

Ihaka, R. and R. Gentleman, R: A language for data analysis and graphics, *J. Computational Graphical Statistics,* 5:299–314, 1996.

Irizarry, R.A. et al., Exploration, normalization, and summaries of high-density oligonucleotide array probe level data, *Biostatistics,* (in press), 2003.

Iyer, V.R. et al., The transcriptional program in the response of human fibroblasts to serum, *Science,* 283:83–87, 1999.

Jain, A.N. et al., Fully automatic quantification of microarray image data, *Genome Res.,* 12:325–332, http:/ /jainlab.ucsf.edu/Projects.html, 2002.

Jain, A.N. et al., Quantitative analysis of chromosomal CGH in human breast tumors associates copy number abnormalities with p53 status and patient survival, *Proc. Natl. Acad. Sci.,* 98:7952–7957, 2001.

Jin, W. et al., The contributions of sex, genotype and age to transcriptional variance in drosophila melanogaster, *Nature Genetics,* 29:389–395, December 2001.

Kato, M., T. Tsunoda, and T. Takagi, Inferring genetic networks from DNA microarray data by multiple regression analysis, *Genome Informatics,* 11:118–128, 2000.

Kaufman, L. and P. Rousseeuw, *Finding Groups in Data: An Introduction to Cluster Analysis,* New York: John Wiley & Sons, 1990.

Kawai, J. et al., Functional annotation of a full-length mouse cdna collection, *Nature,* 409:685–690, February 2001.

Keles, S., M. van der laan, and M.B. Eisen, Identification of regulatory elements using a feature selection method, *Bioinformatics,* 18:1167–1175, 2002.

Kerr, M.K. and G.A. Churchill, Bootstrapping cluster analysis: Assessing the reliability of conclusions from microarray experiments, *Proc. Nat. Acad. Sci.,* 98:8961–8965, 2001a.

Kerr, M.K. and G.A. Churchill, Experimental design for gene expression microarrays, *Biostatistics,* 2:183–201, 2001b.

Kerr, M.K., M. Martin, and G.A. Churchill, Analysis of variance for gene expression microarray data, *J. Computational Biol.,* 7:819–837, 2000.

Kerr, M.K., P. Leiter, and G.A. Churchill, Analysis of a designed microarray experiment, in *Proc. IEEE-Eurasip Nonlinear Signal and Image Processing Workshop,* June 2001.

Khan, J. et al., Classification and diagnostic prediction of cancers using gene expression profiling and artificial neural networks, *Nature Medicine,* 7:673–679, 2001.

Kim, S. et al., General nonlinear framework for the analysis of gene interaction via multivariate expression arrays, *J. Biomedical Optics,* 5(4):411–424, 2000a.

Kim, S. et al., Multivariate measurement of gene expression relationships, *Genomics,* 67:201–209, 2000b.

Klevecz, R.R., Dynamic architecture of the yeast cell cycle uncovered by wavelet decomposition of expression microarray data, *Funct. Integr. Genomics,* 1(2):186–192, 2000.

Klevecz, R.R. and H.B. Dowse, Tuning in the transcriptome: basins of attraction in the yeast cell cycle, *Cell Proliferation,* 33:209–218, 2000.

Kohonen, T., The self-organizing map, *Proc. IEEE,* 78:1464–1479, 1990.

Langmead, C.J. et al., Phase-independent rhythmic analysis of genome-wide expression patterns, in *Proc. Sixth Annu. Int. Conf. on Computational Molecular Biol.,* Washington, D.C., 2002.

Lazzeroni, L. and A. Owen, Plaid models for gene expression data, *Statistica Sinica,* 12:61–86, 2002.

Lee, M.T. et al., Importance of replication in microarray gene expression studies: statistical methods and evidence from repetitive cDNA hybridizations, *Proc. Natl. Acad. Sci.,* 97:9834–9839, 2000.

Lemon, W.J. et al., Theoretical and experimental comparison of gene expression estimators for oligonucleotide arrays, *Bioinformatics,* 18(11):1470–1476, 2002.

Li, C. and W.H. Wong, Model-based analysis of oligonucleotide arrays: expression index computation and outlier detection. *Proc. Natl. Acad. Sci.,* 98:31–36, 2001a.

Li, C. and W.H. Wong, Model-based analysis of oligonucleotide arrays: model validation, design issues and standard error application, *Genome Biol.,* 2(8): research 0032.1–0032.11, 2001b.

Lim, T.-S., W.-Y. Loh, and Y.-S. Shih, A comparison of prediction accuracy, complexity, and training time of thirty-three old and new classification algorithms, *Machine Learning,* 40:203–229, 2000.

Lin, D.M. et al., A spatial map of gene expression in the olfactory bulb, submitted, 2002.

Lipshutz, R.J. et al., High-density synthetic ologonucleotide arrays. *Nature Genetics,* suppl. 21:20–24, 1999.

Liu, X.S., D.L. Brutlag, and J.S. Liu, An algorithm for finding protein DNA binding sites with applications to chromatin-immunoprecipitation microarray experiments, *Nat. Biotechnol.,* 20(8):835–839, 2002.

Lockhart, D. et al., Expression monitoring by hybridisation to high-density oligonucleotide arrays, *Nat. Biotechnol.,* 14:1675–1680, 1996.

Long, A.D. et al., Improved statistical inference from DNA microarray data using analysis of variance and a Bayesian statistical framework, *J. Biological Chemistry,* 276(23):19937–19944, 2001.

Lönnstedt, I. et al., Microarray analysis of two interacting treatments: a linear model and trends in expression over time, submitted, 2002.

Lönnstedt, I. and T.P. Speed, Replicated microarray data, *Statistica Sinica,* 12(1):31–46, 2001.

Macnaughton-Smith, P. et al., Dissimilarity analysis: a new technique of hierarchical subdivision, *Nature,* 202:1034–1035, 1965.

Manduchi, E. et al., Generation of patterns from gene expression data by assigning confidence to differentially expressed genes, *Bioinformatics,* 16:685–698, July 2000.

Marazzi, A., *Algorithms, Routines and S Functions for Robust Statistics,* Wadsworth & Brooks/Cols, 1993.

Mardia, K.V., J.T. Kent, and J.M. Bibby. *Multivariate Analysis*, San Diego, CA: Academic Press, Inc., 1979.

Marton, M.J. et al., Drug validation and identification of secondary drug target effects using DNA microarrays, *Nature Med.,* 4:1293–1301, 1998.

McLachlan, G.J., *Discriminant Analysis and Statistical Pattern Recognition,* New York: John Wiley & Sons, 1992.

McLachlan, G., R. Bean, and D. Peel, D., A mixture model-based approach to the clustering of microarray expression data, *Bioinformatics,* 413–22, 2002.

Moler, E.J., M.L. Chow, and I.S. Mian, Analysis of molecular profile data using generative and discriminative methods, *Physiological Genomics,* 4:109–126, 2000.

Naef, F. et al., From features to expression: high-density oligonucleotide arrays analysis revisited, *Proc. DIMACS Workshop on Analysis of Gene Expression Data,* October 24–26, 2001.

Newton, M.A. et al., On differential variability of expression ratios: improving statistical inference about gene expression changes from microarray data, *J. Computational Biol.,* 8:37–52, 2000.

Pan, W., A comparative review of statistical methods for discovering differentially expressed genes in replicated microarray experiments, *Bioinformatics,* 18(4):546–545, 2002.

Pan, W., L. Lin, and C. Le, Model-based cluster analysis of microarray gene expression data, *Genome Biol.,* 0009.1–0009.8, 2002.

Park, P.J., M. Pagano, and M. Bonetti, A nonparametric scoring algorithm for identifying informative genes from microarray data, *Pacific Symp. on Biocomputing,* 6:52–63, 2001.

Parmigiani, G. et al., *The Analysis of Gene Expression Data: Methods and Software*, New York, Springer, 2003.

Pearson, E. and J. Wishart, Eds., *Gosset, William Sealy "Student's" collected papers,* Cambridge, 1958, published for the Biometrika Trustees University Press.

Perou, C.M. et al., Distinctive gene expression patterns in human mammary epithelial cells and breast cancers, *Proc. Natl. Acad. Sci.,* 96:9212–9217, 1999.

Pilpel, Y., P. Sudarsanam, and G.M. Church, Identifying regulatory networks by combinatorial analysis of promoter elements, *Nature Genetics,* 29(2):153–159, 2001.

Pollack, J.R. et al., Genome-wide analysis of DNA copy-number changes using cDNA microarrays, *Nature Genetics,* 23:41–46, 1999.

Pollard, K.S. and M.J. van der Laan, A method to identify significant clusters in gene expression data, Technical Report 107, Division of Biostatistics, University of California, Berkeley, April 2002.

Pomeroy, S.L. et al., Prediction of central nervous system embryonal tumour outcome based on gene expression, *Nature,* 415(24):436–442 (and supplementary information), 2002.

Press, S.J., *Applied Multivariate Analysis,* New York: Holt, Rinehart & Winston, 1972.

Raychaudhuri, S., J.M. Stuart, and R.B. Altman, Principal components analysis to summarize microarray experiments: application to sporulation time series. *Pac. Symp. Biocomput.,* 455–466, 2000.

Ripley, B.D., *Pattern Recognition and Neural Networks,* New York: Cambridge University Press, 1996.

Roberts, C.J. et al., Signaling and circuitry of multiple mapk pathways revealed by a matrix of global gene expression, *Science,* 287:873–880, 2000. Web supplement.

Rocke, D.M. and B. Durbin, A model for measurement error for gene expression arrays, *J. Computational Biol.,* 8(6):557–570, 2001.

Rosenblatt, F. The perceptron — a perceiving and recognizing automation, Technical Report 86-460-1, Cornell Aeronautical Laboratory, 1957.

Ross, D.T. et al., Systematic variation in gene expression patterns in human cancer cell lines, *Nature Genetics,* 24:227–234, 2000.

Rouillard, J.M., C. J. Herbert, and M. Zuker, Oligarray: genome-scale oligonucleotide design for microarrays, *Bioinformatics,* 18:486–487, 2001.

Ruczinski, I., C. Kooperberg, and M. LeBlanc, Logic regression, *J. Computational Graphical Statistics,* in press http:// biosunOl.biostat.jhsph.edu/-iruczins/html/publications.html.

Saban, M.R. et al., Time course of lps-induced gene expression in a mouse model of genitourinary inflammation, *Physiol. Genomics,* 5:147–160, 2001.

Sapir, M. and G.A. Churchill, Estimating the posterior probability of differential gene expression from microarray data (poster), The Jackson Laboratory, 2000. http://www.jax.org/research/churchill/.

Sasik, R. et al., Extracting transcriptional events from temporal gene expression patterns during dictyostelium development, *Bioinformatics,* 18(1):61–66, 2002.

Schadt, E.E. et al., Analyzing high-density oligonucleotide gene expression array data, *J. Cellular Biochem.,* 80:192–202, 2001a.

Schadt, E.E. et al., Feature extraction and normalization algorithm for high-density oligonucleotide gene expression array data, *J. Cellular Biochem,.* 84(S37):120–125, 2001b.

Schena, M. et al., Quantitative monitoring of gene expression patterns with a complementary DNA microarray, *Science,* 270(5235):467–70, 1995.

Schena, M., Ed., *Microarray Biochip Technology,* Eaton, 2000.

Schena, M. et al., Parallel human genome analysis: microarray-based expression monitoring of 1000 genes, *Proc. Natl. Acad. Sci.,* 93:10614–10619, 1996.

Sharan, R. and R. Shamir, CLICK: A clustering algorithm with applications to gene expression analysis, in *Proc. 8th Int. Conf. on Intelligent Systems for Molecular Biol. (IS MB),* 307–316, 2000.

Shedden, K. and S. Cooper, Analysis of cell-cycle-specific gene expression in human cells as determined by microarrays and double-thymidine block synchronization, *Proc. Natl. Acad. Sci.,* 99(7):4379–4384, 2002.

Sherlock, G., *A Tutorial for Clustering with XCluster,* Stanford, CA: Stanford University. Available online at http://genome–www.stanford.edu/∼sherlock/tutorial.html, 1999.

Shieh, G.S., Y.C. Jiang, and Y.S. Shih, Comparison of support vector machines to some classifiers using gene expression data, 2002.

Sørlie, T. et al., Gene expression patterns of breast carcinomas distinguish tumor subclasses with clinical implications, *Proc. Natl. Acad. Sci..* 98(19):10869–10874. Supplemental information online at http://genome-www.stanford.edu/, 2001.

Speed, T.P. and Y.H. Yang, Direct and indirect hybridizations for cdna microarray experiments, *Sankhya Series A,* 64(3):707–721, 2002.

Spellman, P.T. et al., Comprehensive identification of cell cycle-regulated genes of the yeast saccharomyces cerevisiae by microarray hybridization, *Molecular Biol. of the Cell,* 9:3273–3297, 1998.

Staunton, J.E. et al., Chemosensitivity prediction by transcriptional profiling, *Proc. Natl. Acad. Sci.,* 98:10787–10792, 2001.

Storey, J.D., A direct approach to false discovery rates, *J. R. Statistical Soc., Series B,* 64:474–498, 2002.

Storey, J.D., The positive false discovery rate: A Bayesian interpretation and the q-value. Technical Report 12, Department of Statistics, Stanford University, 2001.

Storey, J.D. and R. Tibshirani, Estimating false discovery rates under dependence, with applications to DNA microarrays, Technical Report 28, Department of Statistics, Stanford University, Stanford, CA, 2001.

Storey, J.D., J. E. Taylor, and D. Siegmund, A unified estimation approach to false discovery rates, Technical Report 623, Department of Statistics, University of California, Berkeley, 2002.

Struyf, A., M. Hubert, and P.J. Rousseeuw, Integrating robust clustering techniques in S-PLUS, *Computational Statistics Data Analysis,* 26:17–37, 1997.

Tamayo, P. et al., Interpreting patterns of gene expression with self-organizing maps: methods and application to hematopoietic differentiation, *Proc. Natl. Acad. Sci.,* 96, 2907–2912, 1999.

Tang, C. et al., Interrelated two–way clustering: An unsupervised approach for gene expression data analysis, in *Proc. 2nd IEEE Int. Symp. on Bioinformatics and Bioengineering,* Bethesda, MD, 2001.

Theilhaber, J. et al., Bayesian estimation of fold-changes in the analysis of gene expression: the pfold algorithm, *J. Computational Biol.,* 8(6):585–614, 2001.

Thomas, J.G. et al., An efficient and robust statistical modeling approach to discover differentially expressed genes using genomic expression profiles, *Genome Res.,* 11:1227–1236, 2001.

Tibshirani, R., G. Walther, and T. Hastie, Estimating the number of clusters in a dataset via the gap statistic, *J. R. Statist. Soco. B.,* 32(2):411–423, 2001.

Tibshirani, R. et al., Diagnosis of multiple cancer types by shrunken centroids of gene expression, *Proc. Natl. Acad. Sci.,* 99(10):6567–6572, 2002.

Tibshirani, R. et al., Clustering methods for the analysis of DNA microarray data, Technical Report, Department of Health Research and Policy, Stanford University, Stanford, CA, 1999.

Troyanskaya, O. et al., Missing value estimation methods for DNA microarrays, *Bioinjormatics,* 17(6):520–525, 2001.

Tryon, R. and D. Bailey, *Cluster Analysis,* New York: McGraw-Hill, 1970.

Tseng, G.C. et al., Issues in cDNA microarray analysis: quality filtering, channel normalization, models of variations and assessment of gene effects, *Nucleic Acids Res.,* 29(12):2549–2557, 2001.

Tusher, V.G., R. Tibshirani, and G. Chu, Significance analysis of microarrays applied to transcriptional responses to ionizing radiation, *Proc. Natl. Acad. Sci.,* 98:5116–5121, 2001.

van der Laan, M.J. and J. Bryan, Gene expression analysis with the parametric bootstrap, *Biostatistics,* 2(4):445–461,2001.

Vapnik, V., Estimation of dependences based on empirical data (in russian). *Nauka, Moskow,* 1979.

Vapnik, V., *Statistical Learning Theory,* 1st ed., New York: John Wiley & Sons, 1998.

Venables, W.N. and B.D. Ripley, *Modern Applied Statistics with S-PLUS,* 4th ed., New York: Springer-Verlag, 2002.

Wallace, D., The Behrens–Fisher and Fieller–Creasy Problems, in *Lecture Notes in Statistics 1, R.A.Fisher: An Appreciation,* Fienberg S.E. and D.V. Hinkley, Eds., New York: Springer-Verlag, 119–147, 1988.

Wen, X. et al., Large-scale temporal gene expression mapping of central nervous system development, *Proc. Natl. Acad. Sci.,* 95:334–339, 1998.

West, M. et al., Predicting the clinical status of human breast cancer using gene expression profiles, *Proc. Natl. Acad. Sci.,* 98:11462–11467, 2001.

Westfall, P.H. and S.S. Young, *Resampling-Based Multiple Testing: Examples and Methods for p-Value Adjustment,* New York: John Wiley & Sons, 1993.

Weston, J. et al., Feature selection for SVMs, *Advances Neural Process. Syst.,*13:668–674, 2001.

Wilhelm, M. et al., Array-based comparative genomic hybridization for the differential diagnosis of renal cell cancer, *Cancer Res.,*62:957–960,2002.

Wodicka, L. et al., Genome-wide expression monitoring in Saccharomyces cerevisiae, *Nat. Biotechnol.,* 15:1359–1367, 1997.

Wolfinger, R.D. et al., Assessing gene significance from cDNA microarray expression data via mixed models, *J. Computational Biol.,* 8(6):625–638, 2001.

Xu, X., J. Olson, and L.P. Zhao, A regression-based method to identify differentially expressed genes in microarray time course studies and its application in an inducible Huntington's disease transgenic model, *Human Molecular Genetics,* 11(17):1977–1985, 2002.

Yang, Y.H. and M.J. Buckley, Spot, http://experimental.act.cmis.csiro.au/spot/index.php, 2001.

Yang, Y.H. and T. Speed, Design issues for cDNA microarray experiments, *Nature Rev. Genetics,* 3(8):579–588, 2002.

Yang, Y.H. et al., Comparison of methods for image analysis on cDNA microarray data, *J. Computational Graphical Statistics,* 11(1):108–136, 2002a.

Yang, Y.H. et al., Normalization for cDNA microarray data, in *Microarrays: Optical Technologies and Informatics, vol. 4266, Proc. SPIE,* Bittner, M.L. et al., Eds., 141–152, May 2001.

Yang, Y.H. et al., Normalization for cDNA microarray data: a robust composite method addressing single and multiple slide systematic variation, *Nucleic Acids Res.,* 30(4):e15, 2002b.

Yates, F., *The Design and Analysis of Factorial Experiments,* Technical Communication No. 35 of the Commonwealth Bureau of Soils, Commonwealth Agricultural Bureaux, 1937.

Yeung, K.Y. et al., Model-based clustering and data transformations for gene expression data, *Bioinformatics,* 17(10):977–987, 2001.

Youden, W.J., Measurement agreement comparisons, in *Precision Measurement and Calibration: Statistical Concepts and Procedures, Volume 1 of Special Publication 300,* H.H. Ku, Ed., National Bureau of Standards, United States Department of Commerce, Washington, D.C., 146–151, February 1969.

Zhang, L. et al., A new algorithm for analysis of oligonucleotide arrays: application to expression profiling in mouse brain regions. *J. Mol. Biol.,* 317:225–235, 2002.

Zhao, L.P., R. Prentice, and L. Breeden, Statistical modeling of large microarray data sets to identify stimulus-response profiles. *Proc. Natl. Acad. Sci.,* 98(10):5631–5636, 2001.

Zhou, Y. and R. Abagyan, Match-only integral distribution (MOID) algorithm for high-density oligonucleotide array analysis BMC, *Bioinformatics,* 3:3, 2002.

Zhu, G. et al., Two yeast forkhead genes regulate the cell cycle and pseudohyphal growth, *Nature,* 406:90–94, 2000.

INDEX

D

E

F

G

H

I

J

L

M

P

Q

R